T0140403

Sustained Simulation Performance 2021

Michael M. Resch • Johannes Gebert
Hiroaki Kobayashi • Wolfgang Bez
Editors

Sustained Simulation Performance 2021

Proceedings of the Joint Workshop
on Sustained Simulation Performance,
University of Stuttgart (HLRS)
and Tohuku University, 2021

 Springer

Editors
Michael M. Resch
High-Performance Computing Center
University of Stuttgart, HLRS
Stuttgart, Germany

Johannes Gebert
High-Performance Computing Center
University of Stuttgart
Stuttgart, Germany

Hiroaki Kobayashi
Graduate School of Information Sciences
Tohoku University
Aoba-ku, Japan

Wolfgang Bez
NEC High Performance Computing
Europe GmbH
Düsseldorf, Germany

ISBN 978-3-031-18048-4   ISBN 978-3-031-18046-0   (eBook)
https://doi.org/10.1007/978-3-031-18046-0

Mathematics Subject Classification (2020): 65-XX, 65Exx, 65Fxx, 65Kxx, 68-XX, 68Mxx, 68Uxx, 68Wxx, 70-XX, 70Fxx, 70Gxx, 76-XX, 76Fxx, 76Mxx, 92-XX, 92Cxx

This Springer imprint is published by the registered company Springer Nature Switzerland AG
The registered company address is: Gewerbestrasse 11, 6330 Cham, Switzerland

Preface

The Workshop on Sustained Simulation Performance was held online at HLRS in March 2021 and in a hybrid mode at the Cyberscience Center, Tohoku University in December 2021. The collaboration between the High-Performance Computing Center Stuttgart, Tohoku University and NEC has been marked by the Covid pandemic, in which we demonstrated our ability to adapt to new situations and continue our partnership. Ultimately, we are happy to continue the relationship that began in 2004 with the establishment of what we called the 'Teraflop Workshop'. While the homepage still remembers this name, the workshop evolved into the Workshop on Sustained Simulation Performance with more than 30 events on two continents.

Perhaps we were able to adapt so quickly to the pandemic because the field of high-performance computing has always evolved rapidly. While HPC systems were designed for many years as single processor vector machines, they now are large cluster systems with fast interconnects and rather typically with a combination of a variety of processors and accelerators – among them still vector processors. Climate and weather simulation is one of the scientific fields that has a particularly high demand for computing power, and research has shown that we want to use our resources more sustainably. This is at odds with the ever larger systems with ever higher energy consumption of modern HPC systems. At the same time, however, there has been a tremendous increase in efficiency. The contributions of this book and the upcoming workshops will help to continue and accelerate the development of fast and efficient high-performance computing.

We would like to thank all the contributors and organizers of this book and the Sustained Simulation Performance Workshops. We especially thank Prof. Hiroaki Kobayashi for his close collaboration over the past years and look forward to intensifying our cooperation in the future.

Stuttgart, Germany
December 2021

Michael M. Resch
Johannes Gebert

Contents

Supercomputer for Quest to Unsolved Interdisciplinary Datascience (SQUID) and its Five Challenges 1
Susumu Date, Yoshiyuki Kido, Yuki Katsuura, Yuki Teramae and Shinichiro Kigoshi

Simulating Molecular Docking on the SX-Aurora TSUBASA Vector Engine 21
Leonardo Solis-Vasquez, Erich Focht and Andreas Koch

Simulation of Field-induced Chiral Phenomena in Inhomogeneous Superconductivity 37
Hirono Kaneyasu, Kouki Otsuka, Singo Haruna, Shinji Yoshida and Susumu Date

Exploiting Hybrid Parallelism in the LBM Implementation Musubi on Hawk 53
Harald Klimach, Kannan Masilamani and Sabine Roller

MPI Continuations And How To Invoke Them 67
Joseph Schuchart and George Bosilca

Xevolver for Performance Tuning of C Programs 85
Hiroyuki Takizawa, Shunpei Sugawara, Yoichi Shimomura, Keichi Takahashi and Ryusuke Egawa

Scalability Evaluation of the CFD Solver CODA on the AMD Naples Architecture 95
Michael Wagner

Supercomputer for Quest to Unsolved Interdisciplinary Datascience (SQUID) and its Five Challenges

Susumu Date, Yoshiyuki Kido, Yuki Katsuura, Yuki Teramae and Shinichiro Kigoshi

Abstract The Cybermedia Center at Osaka University started the operation of a supercomputing system named Supercomputer for Quest to Unsolved Interdisciplinary Datascience (SQUID) in May 2021. SQUID is a hybrid supercomputing system composed of three kinds of heterogeneous compute nodes and delivers 16.591 PFlops as the theoretical performance. This paper overviews the architecture and structure of SQUID and then explains the five challenges which we have set in designing SQUID: Tailor-made computing, HPC and HPDA integration, Cloud-interlinked and -synergized, Secure computing environment, and Data aggregation environment. After that, the future issues to be tackled through the actual operation of SQUID are described.

1 Introduction

Recently, the globalization of academic research has been accelerating. It requires the aggregation and integration of computing, data and even human resources. Accompanied with the globalization of academic research, it would become more common and general that researchers and scientists who are with different organizations work together as a team for solving a common scientific problem [4]. This trend is not exceptional in Osaka University but observed worldwide. For the higher productivity in globalized academic research, the information and communication technologies (ICT) would take a role of greater importance. For the reason, the Cybermedia Center at Osaka University (CMC) which is a supercomputing center and in charge of the administration and management of ICT infrastructures including supercomputing

Susumu Date and Yoshiyuki Kido,
Cybermedia Center, Osaka University, Japan, e-mail: date@cmc.osaka-u.ac.jp

Yuki Katsuura, Yuki Teramae and Shinichiro Kigoshi
Department of Information and Communication Technology Services, Osaka University, Japan

© The Author(s), under exclusive license to Springer Nature Switzerland AG 2023
M. M. Resch et al. (eds.), *Sustained Simulation Performance 2021*,
https://doi.org/10.1007/978-3-031-18046-0_1

system for research and education [2], is expected to implement the supercomputing systems well prepared for the rapid expansion and globalization of academic researches.

Furthermore, high performance data analysis (HPDA) has been increasing its importance. Today, many researchers and scientists are enthusiastic about applying data analysis techniques, characterized with keywords such as artificial intelligence (AI), machine learning (ML) and deep learning (DL), to a large amount of data set to solve their scientific problems. Such enthusiasm, expectation and concern to HPDA have triggered the utilization of supercomputing systems by researchers who have never used any supercomputing system so far. As a result, it is expected that newly developed supercomputing systems should accommodate the new computing needs and requirements derived from HPDA as well as traditional high performance computing (HPC).

In the background above, the CMC has developed and installed a new supercomputing system named Supercomputer for Quest to Unsolved Interdisciplinary Datascience (SQUID) [13] in May 2021, in a hope that the new supercomputing system facilitates researchers and scientists who work on researches for the advancement of academia and industries to explore unsolved data scientific problems. For realizing SQUID, we have set the five challenges toward our envisaged next-generation supercomputing systems. In this paper, we briefly introduce SQUID and then explain the five challenges.

This paper is structured as follows. Section 2 briefly introduces the hardware configuration of SQUID. In Section 3 the five challenges set in realizing SQUID are explained. After that, Section 4 describes the issues to be tackled. Section 5 summarizes this paper.

2 Hardware configuration of SQUID

Figure 1 shows the exterior view of SQUID installed at the CMC. This SQUID is a hybrid supercomputing system composed of three different architectures; general-purpose CPU, GPU and vector nodes. All of processors and accelerators deployed on the compute nodes of SQUID are cooled with DLC (direct liquid cooling) for stable operation and high performance delivery purpose. For the parallel filesystem, Lustre-based DDN EXAScaler was adopted to provide users with a single and fast disk image of 20 PB HDD and 1.2 PB NVMe SSD. Mellanox InfiniBand HDR (200 Gbps) was adopted to connect all of three types of compute nodes and the Lustre parallel filesystem (Fig. 2). As the topology, the combinational use of the Dragonfly+ and Fat-tree was adopted. As to the Dragonfly+ topology for CPU nodes, 1520 CPU nodes are divided to three groups (513 nodes, 513 nodes and 494 nodes) and a group is connected to each of other two groups with 95 IB HDR links (19 Tbps). The CPU nodes in each group are connected as the Fat-tree topology to take advantage of full-bisectional bandwidth. On the other hand, the GPU nodes, the vector nodes, the file servers for Lustre filesystem, and other management servers for SQUID are

Fig. 1: Exterior view of SQUID.

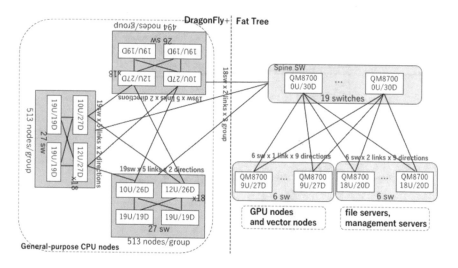

Fig. 2: Overview of the interconnect on SQUID.

connected as the Fat-tree topology to utilize full-bisectional bandwidth. The spine switches of the Fat-tree interconnect for the GPU node, the vector nodes, the file servers and other management servers are connected to each group of the CPU nodes with 36 IB HDR links (7.2 Tbps).

Table 1 shows the system performance and configuration of SQUID. The theoretical performance of SQUID reaches 16.591 PFlops. The major portion of SQUID, as the table indicates, is the cluster of general-purpose CPU nodes. SQUID has 1520

Table 1: System performance and configuration of SQUID.

compute node (16.591 PFlops)	general-purpose CPU nodes 1,520 nodes (8.871 PFlops)	CPU: Intel Xeon Platinum 8368 (Ice Lake / 2.4 GHz 38C) x 2 Memory : 256 GB
	GPU nodes 42 nodes (6.797 PFlops)	CPU: Intel Xeon Platinum 8368 (Ice Lake / 2.4 GHz 38C) x 2 Memory : 512 GB GPU: NVIDIA HGX A100 8-GPU board (Delta)
	vector nodes 36 nodes (0.922 PFlops)	CPU: AMD EPYC 7402P (ROME / 2.8 GHz 24C) x 1 Memory : 128 GB vector: NEC SX-Aurora TSUBASA Type20A x 8
storage	DDN EXAScaler(Lustre)	HDD: 20.0 PB NVMe 1.2 PB
interconnect	Mellanox InfiniBand HDR (200 Gbps)	
front-end node	front-end node for HPC 4 nodes	CPU: Intel Xeon Platinum 8368 (Ice Lake / 2.4 GHz 38C) x 2 Memory : 256 GB
	front-end node for HPDA 4 nodes	CPU: Intel Xeon Platinum 8368 (Ice Lake / 2.4 GHz 38C) x 2 Memory : 512 GB
	secure front-end node 1 node	CPU: Intel Xeon Platinum 8368 (Ice Lake / 2.4 GHz 38C) x 2 Memory : 256 GB

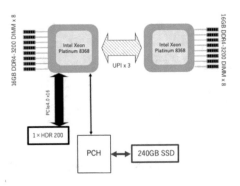

Fig. 3: The internal architecture of a SQUID CPU node.

CPU nodes in total. Figure 3 shows the block diagram of the CPU node. Each CPU node has 2 Intel Xeon Platinum 8368 (Ice Lake/ 2.4 GHz, 38 Core) processors and 256 GB memory deployed. The two processors are connected on 3 UPI (Ultra Path Interconnect) links and 8 channels of 16 GB DDR4-3200 DIMMs are connected to each processor. The memory bandwidth available in each processor is 204.8 GB/s. Each processor delivers 2.918 TFlops as its theoretical performance and so each CPU node delivers 5.836 TFlops. Therefore, the total theoretical performance of CPU nodes on SQUID becomes 8.871 PFlops.

Figure 4 illustrates how CPU nodes are installed. The left is NEC LX103Bj-8, the blade server equipped with two CPU nodes. These 19 blade servers are mounted in the blade enclosure (chassis) shown in the center in the figure. Then, the 4 chassis are mounted to a compute rack. As the result, all of CPU nodes are mounted in 20 compute racks.

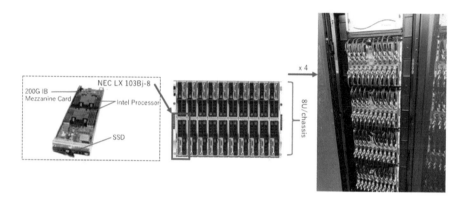

Fig. 4: SQUID CPU nodes.

The second major portion in terms of theoretical performance of SQUID is the cluster of GPU nodes. The total theoretical performance in the cluster of GPU nodes is 6.797 PFlops. The GPU node was designed so that it has the same type of processors as CPU nodes. This reason can be explained from the following experience of our operation of OCTOPUS (Osaka university Cybermedia cenTer Over-Petascale Universal Supercomputer) [9], which is the supercomputing system installed in 2017. Throughout the operation of OCTOPUS, we observe the heavy loaded utilization of general-purpose CPU nodes that have the Intel Xeon processors. Then, we dynamically change the configuration of the scheduler system so that a part of user job requests targeting CPU nodes are transferred to GPU nodes, only when the utilization of GPU nodes are not so high. From this experience, the GPU node was designed to have 2 Intel Xeon Platinum 8368 (Ice Lake/ 2.4GHz, 38 Core) processors to seamlessly accommodate the job requests targeting CPU nodes.

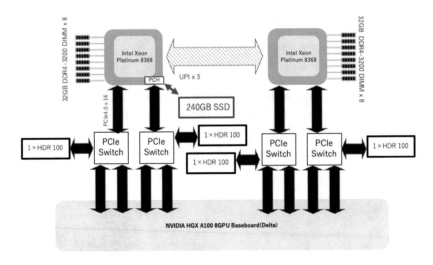

Fig. 5: The internal architecture of a SQUID GPU node.

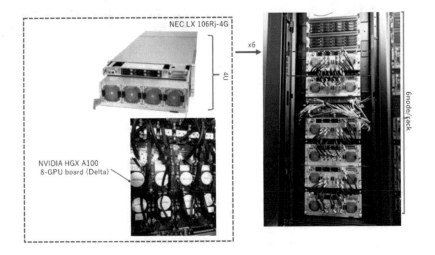

Fig. 6: SQUID GPU nodes.

Figure 5 shows the block diagram of the GPU node. The two processors are connected on 3 UPIs and 8 channels of 32 GB DDR4-3200 DIMMs are connected to each processor. The memory bandwidth in each processor is 204.8 GB/s. Remarkably, the SQUID GPU node has four HDR100 connections to the interconnect, while the CPU node has a single HDR200 connection to the interconnect. The reason is explained from the intention that the data traffic from CPU and GPU are distributed symmetrically in the GPU node. For the GPU accelerator, NVIDIA HGX A100 8-GPU board (Delta) is deployed. It is connected to four PCIe Switches through PCIe4.0x16 links as the figure shows. As the name indicates, eight A100 GPUs are deployed on the board. These A100 GPUs are connected through NVLINK and NVSwitch. In more detail, a single A100 GPU has 12 NVLINK interconnect, each of which supports bi-directional 50 GB/s communication connected to NVSwitch. Therefore, 600 GB/s bi-directional communication can be supported [14]. This A100 GPU delivers 19.5 TFlops (double precision) as the theoretical performance and so each GPU node reaches 161.836 TFlops. Figure 6 shows how GPU nodes are mounted. Six 4U rack-mount servers (NEC LX106Rj-4G) shown in the figure are mounted on a compute rack and thus 7 compute racks in total are installed for SQUID.

The last portion is the cluster of vector nodes. The total theoretical performance of vector nodes is 0.922 PFlops. This vector node is characterized with NEC SX-Aurora TSUBASA Type 20A vector engine which is a vector processor. Figure 7 shows the block diagram of a vector node. It has an AMD EPYC 7402P (2.8 GHz, 24 Core) processor and 128 GB memory as well as 8 NEC SX-Aurora TSUBASA Type 20A vector engines deployed. Eight vector engines are connected to the processor through PCIe Switches on PCIe3.0x16 links. As to the connection to the interconnect, the processor has two PCIe4.0x16 links to two HDR200 HCAs. The vector engine has 10 vector processor cores, each of which delivers 307 GFlops (double precision), and 48 GB HBM2 and thus the peak performance of NEC SX-Aurora TSUBASA Type 20A is 3.07 TFlops. On the other hand, the AMD processor's theoretical performance is 2.15 TFlops. Therefore, the total performance of 36 vector nodes becomes 0.922 PFlops. Figure 8 shows how vector nodes are mounted. Eight NEC SX-Aurora TSUBASA Type20A are deployed onto a 2U rack-mount server. These 18 servers are mounted on a rack. 36 SQUID vector nodes are mounted in 2 racks.

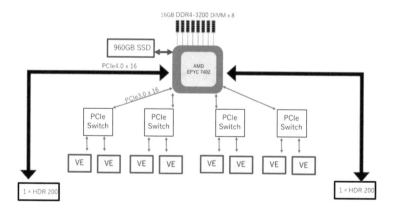

Fig. 7: The internal architecture of a SQUID vector node.

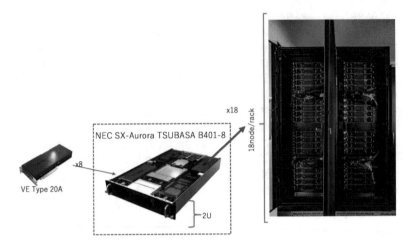

Fig. 8: SQUID vector node.

3 Five challenges behind SQUID

In the procurement of SQUID, we have set the following five challenges that makes SQUID unique and advanced.

1. Tailor-made computing
2. HPC and HPDA integration
3. Cloud-interlinked and -synergized
4. Secure computing environment
5. Data aggregation environment

In this section, these five challenges are explained in terms of the reason why these challenges are aimed and how the challenges are tackled.

3.1 Tailor-made computing

Researchers want to utilize a supercomputing system in their favorite way. For example, some researchers may want to perform their own hand-made program, others may want to utilize open-source software for their analysis. Some researchers want to use either MPI library or OpenMP for parallelization, others both. Even if they use the same software and libraries, the required version of software and libraries give rise a serious problem. Due to the diversification, it is hard and even impossible to prepare a common software stack that all of the users can satisfy. In fact, we encountered this type of problems many times on the supercomputing systems installed in prior to SQUID at the CMC. To solve this problem, we have adopted SingularityCE [12] as container virtualization technology, so that each researcher can make their own software stack or convert a docker container image to the corresponding Singularity image. This challenge is tightly related to HPC and HPDA integration and secure computing environment challenges described later in this paper.

3.2 HPC and HPDA integration

As described in Section 1, the computing needs and requirements in the academic research scene have been diversified. A reason for this diversification can be explained from the newly emerged computing needs and requirements. Many researchers and scientists are fascinated by high performance data analysis characterized by keywords such as AI, ML and DL and have been seeking to apply such HPDA to their scientific and engineering problems. This recent worldwide trend triggers supercomputing centers to consider what the supercomputing system is like in near future. In addition, the fact that there are different types of processors such as Intel, AMD, and ARM and accelerators such as GPU, vector engine and FPGA also illustrates the diversification. Moreover, the number of processor cores differs even in the same type of processors.

The supercomputing centers located in universities are expected to provide supercomputing systems that can smoothly accommodate the computing needs from researchers and scientists necessitating large-scale compute resources. Until recently, the jobs executed on supercomputing systems are mostly numerical analysis and computer simulations that utilize MPI programs written in C or Fortran. Also, these types of traditional jobs accept a batch-job processing. However, new types of computing needs derived from HPDA differ from traditional jobs. For example, such computing needs require Java and Python rather than C and Fortran. Also, researchers and scientists prefer to use supercomputing systems in an interactive manner because researchers who utilize HPDA techniques necessitate trial-and-error workflows for data processing. This fact means that completely different software stacks from traditional HPC environment becomes necessary for HPDA jobs.

The easiest way to satisfy both types of computing needs might be to separate compute resources and then provide the different set of resources for each computing needs. In reality, however, it is too difficult to predict how much computing needs derived from HPC or HPDA domains in designing a supercomputing system. If the prediction is not correct, valuable compute resources can be wasted. Even if the prediction is correct, the usage of the supercomputing system would change during operation. Furthermore, in the recent academic research scene, researchers have started to seek for new computing ways of combining HPC and HPDA techniques. For the reason, the static separation of compute resources in supercomputing systems is not a better way to satisfy both computing needs and requirements.

For the reason above, we explored the integration of HPC and HPDA environments on SQUID so that it can accommodate both types of supercomputing needs. Specifically, we took the strategy to prepare a set of front-end servers designed for HPC users and HPDA users respectively (Table 1), rather than separating the compute nodes of SQUID into two groups. By deploying the software stack targeting HPC or HPDA on each set of front-end servers, we aimed to satisfy the computing needs and requirements from each user group and provide better user experience. In detail, for the HPC users, the corresponding front-end servers provide with a computing environment where users can develop their HPC codes and then submit jobs in the same way as until today. On the other hand, for the HPDA users, the corresponding front-end servers are deployed so that the users can interactively perform their data analysis workflow, for example using Jupyter Notebook, in a trial-and-error manner. Importantly and however, even in the case of using the HPDA front-end servers, our strategy forces the users to wait for compute resources being available when they perform large-scale data analysis jobs. The reason can be explained from the fact that compute resources are finite and our policy that our center would like to provide compute resources fairly for any type of users.

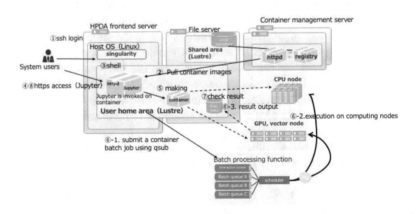

Fig. 9: An example workflow on a HPDA frontend server.

Figure 9 illustrates how the users can utilize SQUID through the use of HPDA frontend servers. The key point in the HPDA front-end server is the use of SingularityCE [12] for providing each user group with a feeling like as if only they are using SQUID. When the user utilizes SQUID for HPDA purpose, the user first logins (ssh) to the HPDA front-end server. After that, the user retrieves the Jupyter Notebook container image and then starts the container. After the container starts, it displays an access URL like https://squidhpda1.hpc.cmc.osaka-u.ac.jp:10125 on the user terminal. Then, the user can perform his/her data analysis interactively, by accessing (https) the access URL from his/her browser and typing his/her user ID and password onto the Jupyter Notebook. If the user wants to perform large-scale data analysis on the compute nodes of SQUID, the user needs to submit a Singularity container job using the *qsub* command to the job scheduler of SQUID.

3.3 Cloud-interlinked and -synergized

In these days a variety of IaaS (Infrastructure as a Service) cloud have prevailed. Representative examples of such cloud include AWS [1], Azure [7] and OCI [11]. The convenience that allows researchers to dynamically build their favorite computing environment through an intuitive interface would be accepted by more researchers in near future. On the other hand, as described, the computing needs and demands have been increasing. From such a perspective, the functionality of using the cloud resources on demand when the computing needs from users exceed the on-premise supercomputing resources is considered necessary in the next-generation supercomputing systems.

From the consideration, we have explored the dynamic use of the cloud resources by introducing a cloud bursting solution onto OCTOPUS, the supercomputing system built in prior to SQUID. As a result, we have indicated that our developed cloud bursting solution was feasible in the paper [3]. Also after that, we have continued to investigate the solution in terms of whether it can be available as a service to our users by providing it with limited number of users on OCTOPUS. After the careful investigation, the cloud bursting solution has been deployed as a product-level functionality onto SQUID.

Figure 10 shows the simplified architecture of the cloud bursting solution deployed between SQUID and Microsoft Azure. Our cloud bursting solution highly respects user convenience. In other words, the solution aims to provide users with a computing environment where users can take advantage of on-premise and cloud compute resources without being aware of their difference. For the purpose, our cloud bursting solution simply extends the internal network through the IPSec VPN (Virtual Private Network) to the cloud and then reinforces the scheduler deployed on SQUID so that the cloud resources are utilized as if they were on-premise resources.

Technically, our cloud bursting solution leverages NQSV [8], which is a NEC proprietary job management system deployed on SQUID. NQSV inherently allows users to submit virtual machine and container jobs. In other words, NQSV enables the invocation of virtual machines and containers as user jobs. By utilizing this functionality, our cloud bursting solution dynamically starts virtual machines on the IaaS cloud when a job request arrives at the cloud bursting queue described below. For the use of filesystem from the cloud resources, the cloud resources mount the on-premise filesystem through the use of NFS on VPN. On SQUID, the similar cloud bursting solution has been deployed between SQUID and OCI although the bare metal servers are used instead of virtual machines in the case of OCI.

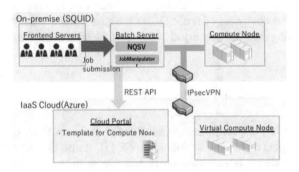

Fig. 10: Cloud bursting solution between SQUID and Microsoft Azure.

At the time of writing this paper, we assume the use of single cloud bursting queue where CPU nodes and the cloud resources on Azure or OCI are mapped for this cloud bursting solution. In this case, the submitted job requests are executed on either the on-premise resources or the cloud resources. In the case that the job request is executed onto the cloud resources, it is easily predicted that the executed job incurs inevitable overhead in performing file I/O operations. From the perspective, we have been working on the realization of intelligent scheduling algorithms that consider computational characteristics for workload distribution [16].

3.4 Secure computing environment

Scientists in the research areas treating security-sensitive data, such as medical and pharmaceutical sciences, have been highly interested in applying high performance data analysis techniques to their own large amount of scientific data. Taking the situation into consideration, the CMC has been working on the realization of the computing environment where scientists and researchers in such research area can utilize supercomputing systems so far [5, 17]. SQUID was designed to provide users with two functionalities for realizing a secure computing environment where security-sensitive data can be treated. The first functionality is the secure partitioning and the second functionality is the secure stating. The secure partitioning functionality allows the user to perform their computation on fully-isolated virtual environment where virtual machines are connected on a virtual network partitioned from the interconnect. The secure staging functionality automatically connects SQUID to a remote storage where confidential data and container images are stored only while security-sensitive data is treated.

This mechanism works as follows. To allow users to utilize a secure computing environment, SQUID provides a secure front-end server (Table 1). When the job requests submitted from the secure front-end server are executed on SQUID, the job's containers are invoked on the compute nodes and then connected to the network logically partitioned from the interconnect. At this time, if the job requests request the secure staging functionality to securely access the data on remote storage, the network between the compute nodes and remote SSD are dynamically established and the container images and data on the remote SSD becomes available.

3.5 Data aggregation environment

As described in Section 1, in the today's academic research scene, researchers and scientists need to collaborate with each other despite the geographical distribution of compute, data and even human resources. In particular, to solve a scientific problem, the data infrastructure facilitating the smooth sharing and exchange of research data has been increasingly more important as well as the computing infrastructure delivering high performance. This tendency is not exceptional in researches using supercomputing systems. Until today the CMC has operated the supercomputing systems to mainly provide a high performance computing service. Most of these supercomputing systems which the CMC has provided were designed and built as an isolated and independent computing infrastructure. The reasons for this isolated computing infrastructure can be explained from the administrators' intention that the administrators want to protect the supercomputing systems so that malicious users cannot attempt to illegally use them. Also, the fact that the administrators want to invest the budget to compute resources as much as possible for pursuing higher computational performance may explain the reasons. However, the globalization in academic research scene now begins to necessitate a data infrastructure that allows them to easily aggregate data to be analyzed and share the analysis result for higher productivity of their research activities. The rising expectation and concern to HPDA has been further increasing this demand.

From the perspective above, we have designed a data infrastructure named Osaka university Next-generation Infrastructure for Open research and open innovatioN (ONION) [10]. Figure 12 shows the overview of ONION. ONION was designed to satisfy the following seven requirements towards the data infrastructure which we envisage.

1. Scientific data generated on diverse data sources such as scientific devices and IoT sensors can be easily accommodated and aggregated.
2. Even researchers who do not have any user account on supercomputing systems can access data on ONION if the users of supercomputing systems want to permit.

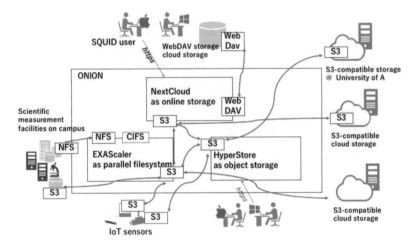

Fig. 11: Overview of ONION.

3. Administration privilege can be delegated to the representative of storage user group and the representative can issue account to each member of the group.
4. Users' external storage can be connected to ONION.
5. ONION and the above user's storage can be viewed as a single disk image to allow users to easily access data of their interest.
6. I/O performance must be high enough to satisfy HPC and HPDA users' requirements.
7. ONION should have the interoperability with GakuNinRDM [6], a research data management infrastructure by National Institute of Informatics (NII), Japan.

As the result of our investigation, we have come up to a conclusion that no single storage solution can satisfy all of these requirements and so we designed ONION by synergically using DDN EXAScaler, Cloudian HyperStore and NextCloud. These three storage solutions were designed to interoperate through the S3 (Amazon Simple Storage Service) protocol. EXAScaler is a Lustre-based parallel filesystem and was adopted for satisfying requirement (6). In addition to DDN EXAScaler, S3DS was adopted to support the S3 protocol. Through the use of S3DS, scientists can access data on EXAScaler even if they do not have user account on SQUID (requirement (2)). Also, requirements (1) and (7) can be easily satisfied.

Cloudian HyperStore is a S3-compatible object storage designed to manage massive amounts of unstructured data. It is a Software-Designed Storage (SDS) platform which runs on any standard x64 server platform. HyperStore supports the S3 protocol and provides higher scalability, meaning that the administrators can flexibly add the size and capacity of storage depending on users' storage needs. Through the use of the S3 protocol, requirements (1), (2) and (7) can be achieved for the same reason as S3DS described above. The most critical reason why we adopted HyperStore is that HyperStore provides multi-tenancy, meaning that it allows the administrator of

HyperStore to delegate a part of administration privilege to each representative of research groups (requirement (7)). In the case of our university, for example, we can delegate the administration privilege to the representative for the Department of Engineering at Osaka University. After that, the representative can make and delete user accounts without obtaining the further administrator privilege depending on the department policy. This delegation functionality reduces the administrator's workload regarding storage management.

NextCloud is an online storage solution and provides an intuitive web user interface for users' data access (Fig. 12). Users can access data through web browsers such as Chrome, Safari and so on. Furthermore, each user can configure his/her NextCloud environment so that local storage can be connected with external storages supporting protocols such as WebDAV and S3 and then the whole disk space can be viewed as a single disk image (requirements (4) and (5)).

Fig. 12: Snapshots of NextCloud on ONION.

These three storage solutions have been integrated so that they interact with one another through the use of the S3 protocol. Also, the external storages on the remote research institutions and organizations can be accessed from ONION and vice versa. Furthermore, S3-compatible IoT sensors can be accommodated on ONION.

4 Future issues towards next-generation infrastructure

The following three issues are currently focused and explored towards our next-generation supercomputing systems.

4.1 Tradeoff between user experience and performance for practical cloud bursting

The cloud bursting functionality on SQUID aims to forward user job requests to the cloud resources without having users being aware of the environmental difference of the on-premise and cloud environments. Therefore, the current implementation of the cloud bursting functionality has sacrificed the achieved performance of user jobs executed on the cloud. As described in Section 3.3, the VPN connection between the cloud and on-premise environments causes the degradation of achieved performance of user jobs executed on the cloud. It is easily predicted that this performance bottleneck takes place under the current implementation, while our implementation enables user job requests to be forwarded without modifying any user code and job script file because the cloud compute nodes are configured to be treated in the same way as the on-premise compute nodes.

To avoid this performance bottleneck, the possible solution would be to prepare a dedicated queue on which only cloud resources are mapped and then let users explicitly submit their jobs to the dedicated queue with the data stating option. This solution is certainly better than our implementation in terms that the jobs executed on the cloud can gain higher performance achieved, particularly when the user jobs are I/O intensive. However, it is harder than the administrators expect for novice users to distinctively submit their jobs to the cloud, by understanding how data stating works, how the job script should be modified from the one targeting the on-premise environment, and so on. In addition to the trade-off problem, we have to control the monetary cost for using the cloud resources for avoiding the shortage of the operational cost of supercomputing systems caused from the overuse of the cloud resource, because the monetary cost is higher than the one for on-premise resources in the CMC. Otherwise, we cannot help temporarily shutting down the supercomputing systems. For the reasons, our operation policy of this cloud bursting functionality is currently to use the cloud resources only when the utilization of compute nodes becomes high and the user waiting time becomes long. Taking the ease of providing the cloud resources with users based on the operation policy into consideration, the cloud bursting queue on our implementation is better with regard to the monetary cost control. For example, this control can be possible just by turning off and on the cloud resources mapped to the cloud bursting queue. Note that the remained jobs in the queue are executed on the on-premise resources after the shutdown of the cloud resources. On the other hand, in the case of the dedicated queue, the users have to wait for the cloud being turned on or resubmit the jobs to the on-premise resources, since the remained queue cannot be executed on the on-premise. Some technical solution that automatically performs data staging on behalf of the job requests submitted to the cloud bursting when the job is executed on the cloud may be a technical issue to solve the above tradeoff problem. Also, the hybrid use of the cloud bursting queue and the dedicated queue might be a practical solution. We have been exploring the better solution to balance user experience and performance through the actual operation of SQUID.

4.2 Software and tools to support integrated way of HPC and HPDA

As described in Section 3.2, researchers have been seeking for new ways of computation. An example of such ways is data assimilation technique. By feeding back the observed data to the simulation, researchers attempt to raise the accuracy of computer simulations. Also, there exists researchers who try to judge whether the simulation computation should be stopped or not by applying AI or ML techniques to the interim result.

However, our current implementation of integrating HPC and HPDA is still primitive and just provides the environment where HPC and HPDA can be performed on SQUID. We have recognized that there is much room for improvement and reinforcement for supporting the exploration for new computational ways by researchers like the above examples. For example, some data mechanism that enables jobs to retrieve data located outside the supercomputing system to the parallel filesystem just before job execution without degrading the job throughput might become necessary for the application of data assimilation technique to simulations executed on the supercomputing system assuming the shared use by multiple research groups [15]. The importance and necessity of such mechanism may be explained also from the recent prevalence of IoT sensors.

In designing ONION, we aimed to realize the seamless data exchange and sharing by researchers. As the result, ONION allows them to easily aggregate such data onto the parallel filesystem and the object storage. Also, it enables researchers to easily share the computational result with the collaborators who does not have any user account on SQUID. However, we consider that the software and tools that allow researchers to benefit from SQUID as the compute infrastructure and ONION as the data infrastructure. Through the operation of SQUID we would like to clarify the requirements and needs by the researchers who explore a new way of computation and then work on the research and development of solutions for such requirements and needs towards the next-generation supercomputing systems.

4.3 Evaluation of secure computing environment for an operational point of view

The secure computing environment realized on SQUID, as described in Section 3.4, provides two functionalities of secure partitioning and secure staging functionality. However, these functionalities still need to be sophisticated from the perspective of performance and stable operation. Currently, the secure partitioning functionality is available on SQUID. However, how much the performance degradation due to the overhead incurred by the dynamic establishment of the secure computing functionality takes place has not sufficiently evaluated. Also, how secure our secure computing environment must be quantitatively evaluated. Otherwise, scientists and researchers cannot trust our secure computing environment for treating security-sensitive data.

For these issues, we plan to collaborate with the dental scientists in Osaka University Dental Hospital through the Social Smart Dental Hospital research collaboration project promoted among the CMC, the hospital and NEC.

5 Summary

In this paper we introduced SQUID, the supercomputing system at the CMC. For making SQUID unique and advanced, we have set five challenges in a hope that SQUID delivers ambitious functionalities for the advancement of globalized academic research. As the result, SQUID not only provides high performance but also offers the functionalities that allow researchers to perform research collaboration with researchers in other universities and research institutions in a highly productive way. For example, by utilizing ONION, which is the data aggregation infrastructure designed for smooth data exchange and sharing, researchers can easily store scientific data to be analyzed on the parallel filesystem and allows their collaborators to download the analysis result immediately after computation. Also, by connecting S3-compatible IoT devices and storages with ONION, each researcher can easily move data between his/her environment and SQUID. For another example, the cloud bursting functionality realized on SQUID enables the administrator to offload the workload to the cloud resources on Azure and OCI in response to the surge in utilization of on-premise compute nodes. This functionality contributes to the reduction of users' job waiting time and results in higher productivity in academic research. Also, the secure computing environment established on SQUID can be used for analysis of security-sensitive data. However, these ambitious features still have to be improved and reinforced through the actual operation. We continue to work on the research and development towards the next-generation supercomputing systems.

References

1. AWS. https://aws.amazon.com/.
2. Cybermedia Center, Osaka University, Japan. https://www.cmc.osaka-u.ac.jp/?lang=en.
3. S. Date, H. Kataoka, S. Gojuki, Y. Katsuura, Y. Teramae and S. Kigoshi. First Experience and Practice of Cloud Bursting Extension to OCTOPUS. In: *Proceedings of 10th International Conference on Cloud Computing and Services Science* (2020). doi: 10.5220/0009573904480455
4. S. Date and S. Shimojo. A Vision Towards Future eScience. In: *Proceedings of 2019 15th International Conference on eScience (eScience)* (2019). doi: 10.1109/eScience.2019.00096.
5. S. Date, T. Yoshikawa, K. Nozaki, Y. Watashiba, Y. Kido, M. Takahashi, M. Muraki and S. Shimojo. Towards A Software Defined Secure Data Staging Mechanism. In: M. Resch, W. Bez, E. Focht, M. Gienger and H. Kobayashi (eds), *Sustained Simulation Performance 2017*, Springer, Cham (2017). doi: 10.1007/978-3-319-66896-3_2
6. GakuNin RDM. https://rcos.nii.ac.jp/en/service/rdm/
7. Microsoft Azure. https://azure.microsoft.com/.

8. NQSV scheduler. https://www.nec.com/en/global/solutions/hpc/articles/tech08.html
9. OCTOPUS. http://www.hpc.cmc.osaka-u.ac.jp/en/octopus/.
10. ONION. http://www.hpc.cmc.osaka-u.ac.jp/en/onion/
11. ORACLE CLOUD Infrastructure. https://www.oracle.com/cloud/.
12. SingularityCE. https://sylabs.io/singularity.
13. SQUID. http://www.hpc.cmc.osaka-u.ac.jp/en/squid/.
14. W. Tsu. Introducing NVIDIA HGX A100: The Most Powerful Accelerated Server Platform for AI and High Performance Computing. https://developer.nvidia.com/blog/introducing-hgx-a100-most-powerful-accelerated-server-platform-for-ai-hpc/.
15. K. Yamamoto, A. Endo and S. Date. Architecture of an On-Time Data Transfer Framework in Cooperation with Scheduler System. In: *Network and Parallel Computing: 18th IFIP WG 10.3 International Conference, NPC 2021, Paris, France, November 3-5, 2021, Proceedings,* Lecture Notes in Computer Science, vol 13152, Springer (2022). doi: 10.1007/978-3-030-93571-9_13
16. S. Yasuda, C. Lee and S. Date. An Adaptive Cloud Bursting Job Scheduler based on Deep Reinforcement Learning. In: *Proceedings of 2021 International Conference on High Performance Big Data and Intelligent Systems (HPBD&IS)* (2021). doi: 10.1109/HPB-DIS53214.2021.9658447
17. T. Yoshikawa et al. Secure Staging System for Highly Confidential Data Built on Reconfigurable Computing Platform. In: *Proceedings of 2019 IEEE International Conference on Computational Science and Engineering (CSE) and IEEE International Conference on Embedded and Ubiquitous Computing (EUC)* (2019). doi: 10.1109/CSE/EUC.2019.00066

Simulating Molecular Docking on the SX-Aurora TSUBASA Vector Engine

Leonardo Solis-Vasquez, Erich Focht and Andreas Koch

Abstract Molecular docking simulations are widely used in computational drug discovery. These simulations aim to predict molecular interactions at close distances by executing compute-intensive calculations. In recent years, the usage of hardware accelerators to speedup such simulations has become essential, since by leveraging their processing capabilities, the time-consuming identification of potential drug candidates can be significantly shortened.

AutoDock is one of the most cited software applications for molecular docking simulations. In this work, we present our experiences in porting and optimizing an OpenCL-based AutoDock implementation on the SX-Aurora TSUBASA Vector Engine. For this purpose, we use device-specific coding techniques in order to leverage the multiple cores on the Vector Engine, as well as its internal vector-based processing capabilities. Based on our experiments, we achieve 3.6× speedup by using a SX-Aurora TSUBASA VE 20B model compared to modern multi-core CPUs, while still achieving competitive performance with respect to modern high-end GPUs.

1 Introduction

Molecular docking simulations are widely used in computational drug discovery. The aim of these simulations is to predict the binding poses between a small molecule and a macromolecular target, both referred to as ligand and receptor, respectively. The purpose of drug discovery is to identify ligands that effectively inhibit the harmful function of a given receptor [5]. In that context, molecular docking simulations play

Leonardo Solis-Vasquez and Andreas Koch
Technical University of Darmstadt, Darmstadt, Germany,
e-mail: solis@esa.tu-darmstadt.de, koch@esa.tu-darmstadt.de

Erich Focht
NEC Deutschland GmbH, Stuttgart, Germany, e-mail: erich.focht@emea.nec.com

a key role at shortening the preliminary identification of potential drug candidates. Subsequent wet lab experiments can be carried out using only a narrowed list of promising ligands, hence reducing the overall cost of experiments.

AutoDock is one of the most cited software applications for molecular docking simulations. It performs an exploration of molecular poses through its main engine: a *Lamarckian Genetic Algorithm* (LGA) [12]. The prediction of the best pose is based on the score, which quantifies the free energy (kcal/mol) of a ligand-receptor system. AutoDock is characterized by nested loops with variable upper bounds and divergent control structures. Moreover, the compute-intensive score evaluations are typically invoked a couple of million of times within each LGA run. Because of this, AutoDock suffers from long execution runtimes, which are mainly attributed to its inability to leverage its embarrassing parallelism. To cope with that, we have recently developed OpenCL-based implementations of AutoDock for speeding up these simulations on a variety of devices including multi-core CPUs and GPUs [19], and even FPGAs [22].

While the aforementioned devices are well-established in the High Performance Computing (HPC) landscape, we think there are other accelerator technologies worth exploring. One of them is the NEC SX-Aurora TSUBASA Vector Engine, whose core technologies are vector-based processing and high-memory bandwidth (1.53 TB/s). The SX-Aurora TSUBASA system offers a productive alternative through a programming framework based on C/C++, and has been recently used to speed up simulations in different fields including computational dynamics, electromagnetism, and others [2, 8, 15].

Moving along these lines, in this work, we present our experiences in developing AutoDock-Aurora, a port and optimization of our OpenCL-based implementation of AutoDock for the SX-Aurora TSUBASA Vector Engine. For achieving higher performance, device-specific coding techniques were applied. We believe that with AutoDock-Aurora, the applicability of the SX-Aurora TSUBASA Vector Engine to solve scientific problems can be expanded.

This paper is organized as follows: Sect. 2 provides a background on molecular docking and on the characteristics of the SX-Aurora TSUBASA Vector Engine. Sect. 3 details how we develop AutoDock-Aurora by describing our porting and optimization experiences. Sect. 4 presents our performance evaluation on the SX-Aurora TSUBASA Vector Engine as well as on CPUs and GPUs. Finally, Sect. 5 presents our closing remarks.

2 Background

This section covers the fundamentals of our work. Sect. 2.1 provides a brief overview of molecular docking and AutoDock's functionality. Sect. 2.2 describes relevant features of the SX-Aurora TSUBASA Vector Engine.

2.1 Molecular docking

Molecular docking explores the poses adopted by a ligand with respect to a receptor. It aims: first, to predict the ligand poses within a certain binding site on the receptor surface; and second, to estimate the binding affinity of their corresponding interactions. A ligand pose can be represented by degrees of freedom experienced during simulation. For instance, the ligand in Fig. 1a possesses three types of degrees of freedom: translation, rotation, and torsion. Typically, such representation involves many degrees of freedom, and thus, results in a molecular docking exploration suffering from a combinatorial explosion. To cope with that, such simulations systematically explore the molecular space using heuristic *search methods* (e.g., genetic algorithms, simulated annealing, etc), which in turn, are assisted by *scoring functions* that estimate the binding affinity [11, 25].

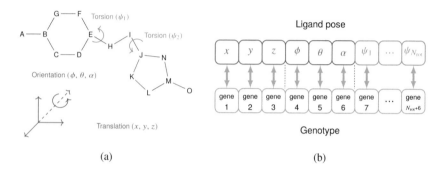

(a) (b)

Fig. 1: (a) Degrees of freedom of a theoretical ligand composed of atoms A, B, C, ..., O. Bonds between atoms are depicted as connecting lines. Each rotatable bond such as E–H and I–J corresponds to a torsion, i.e., rotation of affected ligand atoms around the rotatable-bond axis. (b) Mapping between a ligand pose (a set of degrees of freedom) and a genotype (set of genes). The number of rotatable bonds in a ligand is denoted as N_{rot}

From an algorithmic standpoint, AutoDock executes a Lamarckian Genetic Algorithm (LGA) that combines two methods: a *Genetic Algorithm* and a *Local Search* [12]. The Genetic Algorithm (GA) maps the molecular docking exploration into a biological evolution process. In this mapping, each degree of freedom corresponds to a *gene*. The full set of genes conforms a *genotype* (Fig. 1b), which represents an *individual* of a population. A ligand pose is mapped into an individual, which throughout the evolution, experiences genetic modifications (e.g., crossover, mutation). Moreover, individuals undergo a selection procedure in which the stronger ones survive to the next generation.

The individual's strength, i.e., the score of the respective ligand pose, is estimated with the scoring function. The score quantifies the binding affinity (kcal/mol) by taking into account atomic interactions (Van der Waals, hydrogen bonding, electrostatics, desolvation) and loss of ligand entropy upon binding [6].

The Local Search (LS) further optimizes the scores of the poses already generated via the Genetic Algorithm. For that purpose, during the Local Search execution, AutoDock subjects a population *subset* of randomly-chosen individuals to the Solis–Wets method [20], which aims to minimize the score by performing a number of adaptive iterations. The Solis–Wets algorithm (Fig. 2) takes a genotype as input, and generates a new one by adding small changes (constrained random amount) to each input gene. The scores of the aforementioned genotypes are calculated and compared. In case the score is not minimized, a second genotype is generated by subtracting (instead of adding) small changes to each input gene. Similarly, this is followed by a corresponding score calculation and comparison. The termination criterion of the Local Search is adapted at runtime according to the number of successful or failed score-minimization attempts. The poses improved by the Local Search are re-introduced into the LGA population.

```
1    while ((it < it_MAX) && (step > step_MIN)) {
2        ...
3
4        // Updating counts
5        if (score_lower) {
6            // Updating genotype in one direction
7            genotype = newgenotype_1;
8            succ++;
9            fail = 0;
10           direction = positive;
11       } else {
12
13           // Comparing scores of genotypes
14           if (Score (newgenotype_2) < Score (genotype) {
15               direction = positive;
16           }
17
18           if (direction = negative) {
19               succ = 0;
20               fail++;
21               direction = positive;
22           } else {
23               // Updating genotype in the opposite direction
24               genotype = newgenotype_2;
25               succ++;
26               fail = 0;
27               direction = negative;
28           }
29       }
30   }
```

Fig. 2: Pseudo code of the Solis-Wets method employed as Local Search in AutoDock

Figure 3 depicts the functionality of AutoDock, as well as the default values of its LGA parameters. The Genetic Algorithm is parameterized with the ratios of crossover (R_{cross}), mutation (R_{mut}), and selection (R_{sel}). The Local Search termination is controlled by the minimum change step ($step^{MIN}$), as well as the maximum number of iterations ($N_{LS\text{-}iters}^{MAX}$). A docking job consists of the execution of several independent LGA runs ($N_{\text{LGA-runs}}$). Each of these LGA runs optimizes the scores of a population of $N_{\text{pop-size}}$ individuals. A single LGA run finishes its execution when it reaches the maximum number of score evaluations ($N^{\text{MAX}}_{\text{score-evals}}$) or generations ($N^{\text{MAX}}_{\text{gens}}$), whichever comes first. This figure also shows that the evaluation of an individual's score consists of three steps. In Step 2, the generated pose (expressed as a genotype) is transformed into its corresponding atomic coordinates. In Step 3 and Step 4, the calculated atomic coordinates are used to compute the ligand-receptor (intermolecular) and ligand-ligand (intramolecular) interactions. In addition, the complexity of the score evaluation causes the performance bottleneck, as the scoring contributes to more than 90 % of the total execution runtime.

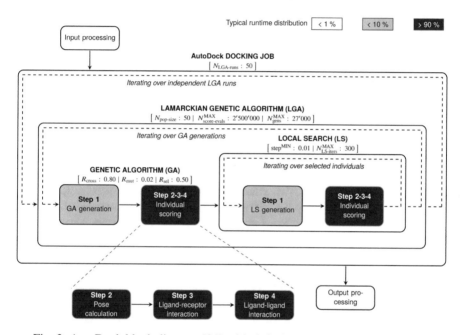

Fig. 3: AutoDock block diagram [21] with default values of LGA parameters

2.2 SX-Aurora TSUBASA Vector Engine

The SX-Aurora TSUBASA Vector Engine, also in this work simply referred to as VE, is a high-performance accelerator in the shape of a full-profile dual-slot PCIe card. It is attached via PCIe to a Vector Host (VH), which is an x86 processor responsible for OS-related tasks as well as for the VE resource management. The VE has eight cores, where each core possesses two processing units. The *scalar processing unit* (SPU) employs a RISC instruction set, out-of-order execution, and

L1 & L2 caches. The *vector processing unit* (VPU) has 64 long vector registers as well as several *vector execution units*. In contrast to conventional SIMD and SIMT architectures, these vector execution units are implemented as 32×64-bit wide SIMD units with 8-cycle deep pipelines. Moreover, the VPU utilizes a *vector length register* and 16 *vector mask registers*. The vector length register controls the number of elements processed in vector operations, while the mask registers enable predication. Currently, VE processors of first (VE 10) and second generation (VE 20) are commercially available [23].

Regarding memory capabilities, the VE features a 48 GB HBM2 RAM, which is capable of up to 1.53 TB/s of memory bandwidth. All eight cores within the VE are connected to a 16 MB Last Level Cache (LLC) through a fast 2D network-on-chip. Moreover, the VE supports two memory-access modes: *normal* and *partitioned*. In the first mode, all VE cores are able to access any part of the LLC and RAM, i.e., they perform uniform memory accesses (UMA). In the second mode, all VE cores are split into two equally-sized groups, and by default, they access only their segment of the LLC and RAM. The usage of the partitioned mode, also known as the non-uniform memory access (NUMA) mode, can result in performance benefits as it reduces the memory-port and memory-network conflicts.

In terms of execution models, the VE supports three different ones [8]: *VE execution*, *VH offload*, and *VE offload*. In the VE execution model, an application is executed on the VE, and only system calls are offloaded to the VH. In the VH offload model, an application is executed on the VE, and system calls and scalar computations are offloaded to the VH. Finally, in the VE offload or accelerator model, the application is executed in the VH, and only compute-intensive calculations (i.e., kernels) are offloaded to the VE. Furthermore, recent studies have developed hybrid programming approaches such as VEDA [24], neoSYCL [7], OpenMP target offloading [1], and HAM [13].

For this work, we use the accelerator model through the programming model called *Vector Engine Offloading* (VEO) [4]. Basically, VEO provides the lowest-level host APIs that can be used to express kernel offloading and VH-VE data movement. Moreover, VEO is based on C++ and its host APIs resemble those of OpenCL.

3 Methodology

This section describes our development of AutoDock-Aurora, and is organized in two parts describing our porting and optimization methodologies.

3.1 Porting

The baseline code of AutoDock-Aurora is ocladock-fpga, an OpenCL implementation of AutoDock tailored for FPGAs [22]. Essentially, we follow the same code partitioning scheme, based on host and device code, already defined in ocladock-fpga. By this scheme, the host takes care of the overall program management, while the device executes the LGA runs offloaded from the host.

Porting for the SX-Aurora TSUBASA Vector Engine requires a proper adaptation of both original host and device codes. For the host code, we adopt the VEO programming model, and thus, we replace the OpenCL API calls with their VEO counterparts. For the device code, porting is much more involved due to its complex implementation in ocladock-fpga, consisting of several OpenCL kernels communicating via OpenCL pipes (i.e., on-chip FIFO-like structures). To adapt this into standard C/C++ functions, we remove all OpenCL-specific language qualifiers, as well as replace OpenCL pipe operations with function calls passing data via pointer arguments.

The porting just described might appear trivial, but it is not due to the implementation complexity of the kernel. In fact, the non-determinism (due to randomness) in the LGA heuristics was the major cause of errors. Therefore, we spent significant development time verifying that the resulting ligand poses and scores reached the expected level of convergence, as discussed in [19].

3.2 Optimization

For the device code, we use the NEC compiler that performs an automatic vectorization. For loops not vectorized in the initial compiler pass, we apply several optimizations (e.g., removal of data dependencies) and code re-factoring (e.g., usage of four-byte int instead of single-byte char for index and loop-control variables). Consequently, we achieved a full vectorization of the functions computing the ligand-receptor (Step 3) and ligand-ligand (Step 4) interactions, which largely contribute to the total execution runtime (Sect. 2.1). Moreover, we leverage the multiple VE cores in the SX-Aurora TSUBASA by distributing the outermost-loop iterations among such cores. This is achieved by annotating the loop executing the independent LGA runs (Fig. 3) with the #pragma omp parallel for compiler directive.

The following subsections organize our optimizations into main and additional ones.

3.2.1 Main Optimizations

At this point in development, although AutoDock-Aurora's code was vectorized and parallelized, it ran ~2.2× slower compared to the host CPU. This low performance was due to the vector pipes being leveraged only for the *innermost* loops, which could be quite short compared to vector length of the VE (= 256 elements, 64-bit each). In particular, the main loops in the score evaluation iterate over: the number of required rotations (Step 2), the number of ligand atoms (Step 3), and the number of atomic pairs contributing to the ligand-ligand interaction (Step 4). For one of the inputs tested, e.g., 1ig3, the aforementioned loop lengths were 53, 21, and 122, respectively. Correspondingly, these loop bounds lead to vector lengths shorter than ¼ th, ¹/₁₂ th, and ½ th of the maximum vector length of the VE.

In order to increase the vector lengths of the device code, we optimize the employed OpenCL-to-VE mapping (Sect. 3.1). In the initial implementation, we mapped each OpenCL thread to a *VE core*. This mapping was replaced by a scheme that instead maps each OpenCL thread to a *vector lane*. In coding terms, we can achieve this by applying the technique called *loop pushing* to the LGA, and thus, to its main components, the Genetic Algorithm (GA) and the Local Search (LS).

Figure 4 shows how we apply loop pushing in the GA code. Basically, the outermost loop iterating over all individuals (i.e., over their genotypes) can be *pushed* into each component of the scoring function (Step 2, Step 3, Step 4), so that this loop becomes the innermost, data parallel, and easily vectorizable loop. For optimal performance, the loop pushing technique is paired with changes in the data layout, so that the vectorized code accesses data with unit-strides as much as possible.

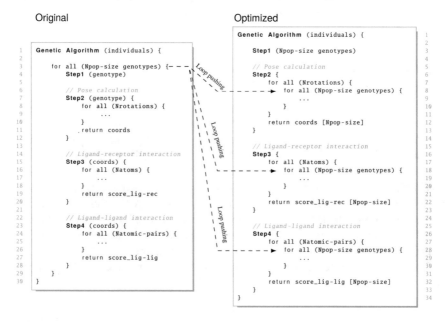

Fig. 4: Optimization in Genetic Algorithm (GA): pushing the outer loop into the three components of the scoring function (Step 2, Step 3, Step 4)

Figure 5 shows the additional techniques required to successfully apply loop pushing in the LS code. The reason for this extra work is the *divergent* nature of the Solis–Wets method implemented as Local Search (Sect. 2.1). In contrast to the GA, where all individuals (of the population) undergo a regular evolution (crossover, mutation, and selection), the individuals in LS can evolve in different directions, or some might converge (i.e., they meet the termination criteria of Solis–Wets) earlier than others. The already-converged individuals are removed from the computation, which in turn, reduces the length of the innermost loop. We are able to apply loop pushing in the LS code by using *predication*, as well as by *compressing* the data of the non-converged part of the population.

3.2.2 Additional Optimizations

The generation of individuals in both GA and LS requires random numbers, which in ocladock-fpga [22] used as baseline, are generated with a congruential random number generator. The disadvantage of such scheme is that each of its generated random values depends on the previous one, i.e., $X_{n+1} = f(X_n)$, which hinders vectorization and parallelization. To cope with this, we replaced the congruential generator with a 64-bit Mersenne Twister pseudo-random generator provided in the NEC NLC library collection [14].

Regarding the numerical precision, we opt to utilize single-precision floating point, similarly as in prior work such as ocladock-fpga [22] and AutoDock-GPU [19]. Moreover, in order to vectorize single-precision computations on the SX-Aurora TSUBASA, we use *packed* vector instructions where each 64-bit vector element of a vector register contains *two* 32-bit float entities. Hence, by enabling *packed vectorization* on top of loop pushing, vectors in device code can have lengths of up to 512 elements, allowing the performance to be doubled.

4 Evaluation

For our experiments, we selected a total of 31 ligand-receptor inputs from [9]. Table 1 shows a subset of ten inputs. Any selected input has less than eight rotatable bonds ($N_{rot} < 8$), as that is the maximum number that can be effectively handled by the Solis–Wets method as Local Search [19].

The following subsections organize our evaluation into execution profiling and performance comparison against other devices.

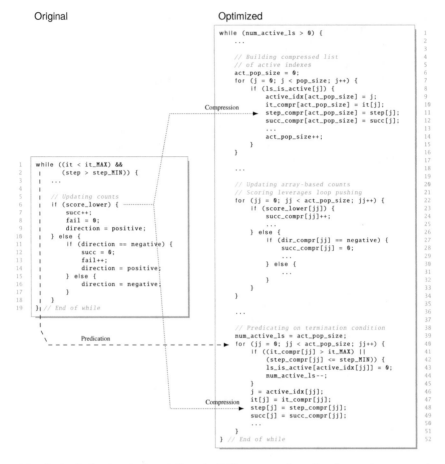

Fig. 5: Optimization in Local Search (LS): usage of predication and compression. In the optimized code, predication updates the number of active individuals. An example of compression-based optimization is the replacement of the succ scalar variable with the succ_compr[ ] array counterpart. In both cases, the number of successful search attempts is counted. In the optimized code, however, the array compresses data for all active individuals

Table 1: Subset of ligand-receptor inputs with their respective number of rotatable bonds (N_{rot}) and atoms (N_{atom})

Input	1ac8	1hnn	1yv3	1owe	1p62	1n46	1ig3	1t46	2bm2	1mzc
N_{rot}	0	2	2	3	4	5	6	6	7	8
N_{atom}	8	18	23	27	22	28	21	40	33	38

Table 2: Execution metrics using the 1ig3 input, before and after applying loop pushing. Information was obtained using NEC PROGINF [18]

Metric	Before	After	Ratio B / A
Real Time [sec]	307.5	14.5	21.3
User Time [sec]	2'458.1	115.0	21.4
Vector Time [sec]	510.2	104.0	4.91
Inst. Count	5'085'000'001'257	98'888'607'313	51.4
Vec. Inst. Count	120'865'697'285	32'136'492'289	3.76
FLOP Count	4'982'577'754'822	4'826'280'301'843	1.03
MOPS	6'012.0	75'174.3	0.08
MOPS (Real)	48'082.1	597'857.0	0.08
MFLOPS	2'027.0	41'960.7	0.048
MFLOPS (Real)	16'211.5	333'711.3	0.049
Avg. Vec. Length	71.5	216.9	0.33
V. Op. Ratio [%]	66.4	99.2	0.67

4.1 Execution profiling

One important aspect to analyze are the performance metrics resulting out of execution profiling. For this purpose, we use the PROGINFO and FTRACE utilities [18], which provide a set of performance counters as well as derived performance metrics. As discussed in Sect. 3.2.1, the major optimization in our work consists in applying loop pushing. Via execution profiling, we can compare relevant execution metrics *before* and *after* applying loop pushing, and thus, analyze the impact of this technique.

In Table 2, the first three are latency metrics. The real time represents the wall-clock elapsed time, while the user time represents the time spent by all eight cores in the VE. As described in Sect. 3.2, the independent LGA runs are distributed among the eight VE cores, and thus, the user time is ~8× that of the real time. The vector time represents the execution runtime for vector instructions. For these latency metrics, we observed reductions (i.e., improvements) of ~21× (real and user time) and ~4.9× (vector time).

The following metrics represent the instruction and operation count. Particularly, the number of all executed instructions, i.e., Inst. Count, is reduced ~51×, while the number of vector instructions is reduced ~3.8×. The reduction in the overall instruction count is attributed to the fact that the initially scalar loops are now vectorized with larger vector lengths (> 200). On the other hand, the number of vector instructions is also reduced because the formerly short loops (i.e., with average vector length of ~72) are now executed as longer loops (i.e., with average vector length of ~217). Moreover, the number of floating-point operations, i.e., FLOP Count, stays almost the *same* in both cases (before vs. after). This is because the program computes the same problem, and hence, it performs roughly the same number of floating-point operations.

Table 3: Technical characteristics of the SX-Aurora TSUBASA VE 20B, GPUs (RTX2070, V100, A100) and CPUs (EPYC ones) used in the evaluation: semiconductor process size (Proc. Size), base clock frequency (Freq), number of cores (Ncores), FP32 performance (Perf), memory bandwidth (MemBW). Both CPU platforms posses two sockets each

Characteristics	VE 20B	RTX2070	V100	A100	EPYC 7502	EPYC 7742
Proc. Size [nm]	16	12	12	7	7	7
Freq [GHz]	1.60	1.61	1.23	0.76	2.50	2.25
Ncores	8	2560	5120	6912	32×2	64×2
Perf [TFLOPS]	4.9	9.1	14.1	19.5	2.6	4.6
MemBW [GB/s]	1530	448	897	1555	204.8×2	204.8×2

The metric termed MOPS represents the number of overall operations per second, while MFLOPS represents the number of floating-point operations per second. The additional metrics, MOPS Real and MFLOPS Real, take into account the eight cores in the VE. Both MOPS and MFLOPS present a significant increase of $12.5\times$ ($= 1/0.08$) and $20.4\times$ ($= 1/0.049$), respectively. Both improvements are due to the increase in: the average vector length (from $\sim$72 up to $\sim$217), and in the vector operation ratio (from 66.4 % up to 99.2 %).

4.2 Performance comparison against CPUs and GPUs

In this section, we compare the execution runtimes achieved on the VE against those achieved on CPUs and GPUs. Table 3 lists the accelerator devices used in our benchmark. In order to run dockings on CPUs and GPUs, we use AutoDock-GPU, the state-of-the-art OpenCL-based implementation of AutoDock. In particular, for ensuring a fair comparison, we use v1.1 of AutoDock-GPU, which features an equivalent functionality to that implemented in AutoDock-Aurora in this work. Note that in Sect. 2.1, we indicated that AutoDock subjects a population *subset* to Local Search. However, in our experiments, we set the *entire* population to undergo Local Search, as in real-world experiments with AutoDock-GPU.

Figure 6 shows the impact of the population size on the execution runtime. We observe that larger population sizes result in faster AutoDock-Aurora executions on the VE. The reason for this are the pushed-in loops that enable longer vector lengths for larger population sizes. On the other hand, we also notice that population sizes do not impact much on the other devices. The different performance behavior on CPUs and GPUs is attributed to the workload distribution employed in AutoDock-GPU. Particularly, AutoDock-GPU spawns a number of OpenCL work-groups that is directly determined by the population size: $N_{\text{WG}} = N_{\text{pop-size}} \times N_{\text{LGA-run}}$. As described in Sect. 2.1, the LGA terminates when the number of score evaluations reaches an upper bound ($N_{\text{score-evals}}^{\text{MAX}}$). Hence, processing larger population sizes requires *fewer* iterations per LGA run, which compensates for the seemingly bigger workload imposed by the need to process more individuals. Moreover, larger population sizes result in additional OpenCL work-groups, which in turn, introduce a synchronization overhead causing the *slight* increase of execution runtimes on CPUs and GPUs.

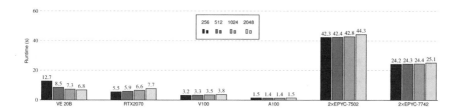

Fig. 6: Geometric mean of execution runtimes over 31 inputs, comparing the impact of the chosen population size: $N_{\text{pop-size}} = \{256, 512, 1024, 2048\}$. AutoDock-Aurora was executed on the VE 20B, while AutoDock-GPU v1.1 on the GPUs and CPUs. In all executions: $N_{\text{LGA-runs}} = 100$. Other parameters were left at default values

For the sake of clarity, Fig. 7 shows only the results when using a population of 2048 individuals ($N_{\text{pop-size}} = 2048$). This configuration is optimal for the VE, but we think there is no disadvantage for other devices when using this configuration for a more detailed comparison. With respect to the CPUs, the VE 20B achieves faster executions than both dual-socket state-of-the-art AMD EPYC nodes by average factors of 6.5× (= 44.3/6.8) and 3.6× (= 25.1/6.8). With respect to the GPUs, executions on the VE 20B are slightly faster than those on the RTX2070 by an average factor of 1.1× (= 7.7/6.8), but slower than those on the V100 and A100 by average factors of 1.8× (= 6.8/3.8) and 4.5× (= 6.8/1.5), respectively. Considering the semiconductor technology used in the fabrication of the selected devices (Table 3), we can put the above factors into perspective: the VE 20B (16 nm) outperforms CPUs that are multiple silicon generations ahead of it (i.e., 7 nm in EPYC), while still achieving competitive performance with respect to GPUs that are at least one generation ahead of it (i.e., 12 nm, 12 nm, 7 nm in RTX2070, V100, A100, respectively).

5 Conclusions

In this work, we have developed AutoDock-Aurora, a port of the AutoDock molecular docking simulation program for the SX-Aurora TSUBASA Vector Engine. Our code baseline was based on OpenCL, and the porting experience was smooth. However, the performance optimization required a combination of a number of device-specific coding techniques. The main technique consisted in increasing the vector lengths via loop pushing, which involved large code-refactoring in the Local Search part of

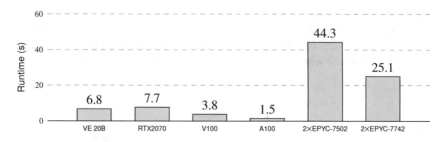

Fig. 7: Geometric mean of execution runtimes over 31 inputs. AutoDock-Aurora was executed on the VE 20B, while AutoDock-GPU v1.1 on the GPUs and CPUs. In all executions: $N_{\text{popsize}} = 2048$, $N_{\text{LGA-runs}} = 100$. Other parameters were left at default values

AutoDock, but in return, reduced significantly the execution runtimes. As a result, AutoDock-Aurora running on a VE 20B is in average 3.6× faster than 128-core CPU servers, while still being competitive to RTX2070, V100, and A100 GPUs.

References

1. T. Cramer, M. Römmer, B. Kosmynin. OpenMP Target Device Offloading for the SX-Aurora TSUBASA Vector Engine. In: *13th Int. Conf. Parallel Processing and Applied Mathematics (PPAM)*, Springer (2019).
2. R. Egawa, S. Fujimoto, T. Yamashita, et al. Exploiting the Potentials of the Second Generation SX-Aurora TSUBASA. In: *Performance Modeling, Benchmarking and Simulation of High Performance Computer Systems (PMBS)*, IEEE (2020).
3. *FightAIDS@Home*. World Community Grid. https://www.worldcommunitygrid.org. Cited01Jun2021
4. E. Focht. VEO and PyVEO: Vector Engine Offloading for the NEC SX-Aurora Tsubasa. In: *Sustained Simulation Performance 2018 and 2019*, Springer (2020).
5. I. Halperin, B. Ma, H. Wolfson, et al. Principles of docking: An overview of search algorithms and a guide to scoring functions. *Proteins: Struct., Funct., Bioinf.* **47**(4), 409–443 (2002). doi: 10.1002/prot.10115
6. R. Huey, G.M. Morris, A. . Olson, et al. A semiempirical free energy force field with charge-based desolvation. *J. Comput. Chem.* **28**(6), 1145–1152 doi: 10.1002/jcc.20634
7. Y. Ke, M. Agung and H. Takizawa. neoSYCL: a SYCL implementation for SX-Aurora TSUBASA. In: *Int. Conf. on High Performance Computing in Asia-Pacific Region (HPC Asia)*, ACM (2021).
8. K. Komatsu, S. Momose, Y. Isobe, et al. Performance Evaluation of a Vector Supercomputer SX-Aurora TSUBASA. In: *Int. Conf. for High Performance Computing, Networking, Storage and Analysis (SC18)*, IEEE (2018).
9. S. LeGrand, A. Scheinberg, A.F. Tillack, et al. GPU-Accelerated Drug Discovery with Docking on the Summit Supercomputer: Porting, Optimization, and Application to COVID-19 Research. *Int. Conf. on Bioinformatics, Computational Biology and Health Informatics (BCB)*, ACM (2020).
10. *List Statistics | TOP500*. Top 500 – The List. https://www.top500.org/statistics/list.Cited01Jun2021

11. J. Liu, R. Wang and J. Chem. Classification of Current Scoring Functions. *J. Chem. Inf. Model.* **55**(3), 475–482 (2015). doi: 10.1021/ci500731a

12. G.M. Morris, D.S. Goodsell, R.S. Halliday, et al. Automated docking using a Lamarckian genetic algorithm and an empirical binding free energy function. *J. Comput. Chem.* 19(14), 1639–1662 (1998). doi: 10.1002/(SICI)1096-987X(19981115)19:14<1639::AID-JCC10>3.0.CO;2-B

13. M. Noack, E. Focht and T. Steinke. Heterogeneous Active Messages for Offloading on the NEC SX-Aurora TSUBASA. In: *Int. Parallel and Distributed Processing Symposium Workshops (IPDPSW)*, IEEE (2019).

14. *Numeric Library Collection 2.3.0 User's Guide*. NEC. https://www.hpc.nec/documents/sdk/SDK_NLC/UsersGuide/main/en/index.html.Cited01Jun2021

15. A. Onodera, K. Komatsu, S. Fujimoto, et al. Optimization of the Himeno Benchmark for SX-Aurora TSUBASA. In: *Benchmarking, Measuring, and Optimizing (Bench)*, Springer (2020).

16. *OpenPandemics: COVID-19*. World Community Grid. https://www.worldcommunitygrid.org/research/opn1/overview.do.Cited01Jun2021

17. N.S. Pagadala, K. Syed and J. Tuszynski. Software for molecular docking: a review. *Biophys. Rev.* **9**(2), 91–102 (2017). doi: 10.1007/s12551-016-0247-1

18. *PROGINF/FTRACE User Guide*. NEC. https://www.hpc.nec/documents/sdk/pdfs/g2at03e-PROGINF_FTRACE_User_Guide_en.pdf.Cited01Jun2021

19. D. Santos-Martins, L. Solis-Vasquez, A.F. Tillack, et al. Accelerating AutoDock4 with GPUs and Gradient-Based Local Search. *J. Chem. Theory Comput.* **17**(2), 1060–1073 (2021). doi: 10.1021/acs.jctc.0c01006

20. F.J. Solis and R.J.B. Wets. Minimization by Random Search Techniques. *Math. Oper. Res.* **6**(1), 1–158 (1981). doi: 10.1287/moor.6.1.19

21. L. Solis-Vasquez and A. Koch. A Performance and Energy Evaluation of OpenCL-accelerated Molecular Docking. In: *5th Int. Workshop on OpenCL (IWOCL)*, ACM (2017).

22. L. Solis-Vasquez and A. Koch. A Case Study in Using OpenCL on FPGAs: Creating an Open-Source Accelerator of the AutoDock Molecular Docking Software. In: *5th Int. Workshop on FPGAs for Software Programmers (FSP)*, VDE Verlag (2018).

23. *Vector Engine Models*. NEC. https://www.nec.com/en/global/solutions/hpc/sx/vector_engine.html.Cited07Mar2022

24. *VEDA GitHub repository*. https://github.com/SX-Aurora/veda.Cited07Mar2022

25. Z. Wang, H. Sun, X. Yao, et al. Comprehensive evaluation of ten docking programs on a diverse set of protein-ligand complexes: the prediction accuracy of sampling power and scoring power. *Phys. Chem. Chem. Phys.* **18**, 12964–12975 (2016). doi: 10.1039/C6CP01555G

Simulation of Field-induced Chiral Phenomena in Inhomogeneous Superconductivity

Hirono Kaneyasu, Kouki Otsuka, Singo Haruna, Shinji Yoshida and Susumu Date

Abstract We explain the field-induced chiral phenomena in inhomogeneous super-conductivity and perform a computational simulation to demonstrate such phenomena on the basis of the Ginzburg–Landau equation for the inhomogeneous interface superconductivity of a eutectic system. Field-induced chiral phenomena occur because of the paramagnetic coupling of an intrinsic magnetization with an external magnetic field. Applying a magnetic field to a non-chiral state leads to a field-induced chiral transition with the generation of a paramagnetic chiral current. Numerically solving the aforementioned equation yields converged solutions and output numerical data obtained through an iterative process. The actual time for this calculation can be distinctly reduced through acceleration via code optimization that is suitable for vector parallelization. Reducing the calculation time makes it possible to extend the simulation to lower temperatures where the inhomogeneous superconductivity spreads to a greater distance from the interface.

Hirono Kaneyasu, Kouki Otsuka and Shingo Haruna
Graduate School of Science, University of Hyogo, Japan, e-mail: hirono@sci.u-hyogo.ac.jp

Shinji Yoshida
Graduate School of Information Science and Technology, Osaka University, Japan

Susumu Date
Cybermedia Center, Osaka University, Osaka, Japan

1 Field-induced chiral phenomena in inhomogeneous superconductivity

1.1 Chiral state

Superconductivity is a valuable feature of energy and electric technologies. Electric properties and electromagnetic features are used in power transmission, magnetic, and quantum devices. The superconducting state appears when the temperature is lowered and two electrons form a pair known as the Cooper pair. Electromagnetic features can be understood from the microscopic aspect of electron pairs [1, 25]. The electron pairs condense in the same quantum state, leading to the macroscopic phenomenon of superconductivity.

The features of a Cooper pair are characterized by the "spin" and "orbital" of the electrons in the quantum state, as shown in Fig. 1 (a). The spin configuration is anti-parallel or parallel, referred to as spin-singlet and spin-triplet states, respectively. Meanwhile, the intrinsic angular momentum L_z for an orbital of a Cooper pair is characterized by an intrinsic magnetization denoted by $L_z \neq 0$, i.e., $L_z = \pm 1, \pm 2, ...,$ which reflects a chiral state with the time-reversal symmetry breaking [17, 32]. The superconducting state with intrinsic magnetization causes interesting chiral phenomena that differ from usual superconductivity.

1.2 Field-induced chiral phenomena

In the chiral state, an electron pair with intrinsic magnetization has a feature response to an external magnetic field because the intrinsic magnetization couples with the external magnetic field [23]. A part of this is the field-induced chiral phenomena, which results from the paramagnetic coupling of the intrinsic magnetization with an external magnetic field [9, 23]. The paramagnetic coupling of the intrinsic magnetization stabilizes the chiral state by generating a paramagnetic chiral current in the direction opposite to the screening current, as shown in Fig. 1 (b) [9, 23]. By contrast, the screening current flows to generate a diamagnetic field to the external magnetic field. This is known as the Meissner effect, which is a general feature of superconductivity [24]. Such field-induced chiral phenomena occur distinctly in the case of inhomogeneous superconductivity, which has been reported in a theoretical study based on the Ginzburg–Landau theory [9].

In inhomogeneous superconductivity, the application of a magnetic field to a non-chiral state causes the field-induced chiral transition with the generation of a paramagnetic chiral current, as shown in Fig. 2 (a) [9]. For example, such a non-chiral state yields an onset temperature of superconductivity as an interface state nucleating near the interface between superconductivity and a metal, which transits to a chiral state when the temperature is lowered in a zero field [9, 11, 12, 31]. Such an interface system is a characteristic of a eutectic superconductor containing multiple interfaces between the parent superconductor and metal inclusions. Identifying the field-induced chiral phenomena in such materials is useful for discovering candidates for chiral superconductors, because the field-induced chiral phenomena constitute evidence of chiral superconductivity.

In this study, a computational simulation is conducted to demonstrate the field-induced chiral phenomena with a paramagnetic chiral current in the inhomogeneous state. In particular, the field-induced chiral transition is produced in an interface superconducting model for eutectic Sr_2RuO_4-Ru [19–21]. The features of the field-induced chiral phenomena are qualitatively compared with the experimental results obtained for eutectic Sr_2RuO_4-Ru [14, 20, 36]. The good agreement with the experimental results serves as evidence of a chiral state in the bulk state of the parent superconductor Sr_2RuO_4 [9].

2 Field-induced chiral phenomena in a eutectic superconductor

2.1 Inhomogeneous interface superconductivity

Sr_2RuO_4 is a potential candidate for chiral superconductors [6, 18, 35]. The parent material of eutectic Sr_2RuO_4-Ru is Sr_2RuO_4, which contains micrometer-scale Ru-metal inclusions [20, 21]. Experiments have reported the nucleation of inhomogeneous superconductivity around interfaces between the Ru-metal inclusions and the parent superconductor Sr_2RuO_4. Inhomogeneous interface superconductivity exhibits an intrinsic magnetization below the bulk transition temperature T_{bulk} near $T_{c,SRO} = 1.5$ K, which is the superconducting transition temperature of pure Sr_2RuO_4 [30]. Moreover, theoretical studies have suggested that the superconducting phenomena are evidence of a chiral state in the parent superconductor Sr_2RuO_4 [3, 9, 11–13, 22, 31].

In the eutectic superconductor Sr_2RuO_4-Ru, the inhomogeneous interface state asymptotes to a superconducting state of pure Sr_2RuO_4 when the temperature is lowered to $T_{c,SRO}$. By contrast, interface superconductivity appears at the onset temperature T_{onset}, i.e., 3 K above T_{bulk}. When the temperature is lowered in a zero field, first, the interface state is a non-chiral state near T_{onset}; thereafter, the non-chiral state transits to a chiral state when T is lowered toward T_{bulk}. This interface superconducting model for explaining the chiral transition is considered for the 3-Kelvin phase model of eutectic Sr_2RuO_4-Ru [31]. The interface superconductivity nucleates in accordance with the increase in the superconducting transition temperature locally near the interface originating from a particular electron state induced by strain due to the deposition of Ru-metal inclusions [7, 33, 37]. This interface model is shown in Fig. 2 (b) [9, 11, 12, 31]. Considering a Ru-metal inclusion in the parent superconductor Sr_2RuO_4, a planar interface perpendicular to a RuO_2-layer is set at the junction Ru/Sr_2RuO_4.

(a)

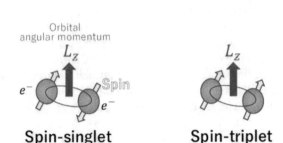

(b)

Inhomogeneous

Paramagntic coupling

Intrinsic magnetization

Magnetic Field

Chiral stability + Paramagnetic supercurrent

Pair breaking + Screening supercurrent

Fig. 1: (a) Chiral state of the Cooper pair in configurations of spin and angular momentum. The intrinsic angular momentum for the orbital of the Cooper pair is $L_z \neq 0$, i.e., $L_z = \pm 1, \pm 2, \dots$. (b) Field-induced chiral phenomena in inhomogeneous superconductivity. The chiral state is stabilized by applying an external magnetic field with the generation of a paramagnetic chiral supercurrent that flows in the direction opposite to the screening supercurrent, whose field is diamagnetic to the external magnetic field.

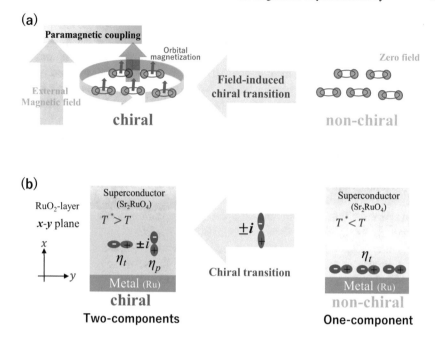

Fig. 2: (a) Field-induced chiral transition in inhomogeneous superconductivity. The chiral state stabilizes owing to the paramagnetic coupling of an intrinsic magnetization with an external magnetic field, and it generates a paramagnetic chiral supercurrent. (b) Order parameter η of inhomogeneous superconductivity near an interface between a metal and a superconductor, i.e., Ru/Sr_2RuO_4. The chiral state is represented by two components η_t and η_p in the xy-plane parallel to the RuO_2-layer, and this is common to the chiral states $k_{zx} \pm ik_{yz}$ and $k_x \pm ik_y$ in the projection to the xy-plane. Here, η_t and η_p denote the tangential and perpendicular components, respectively. A two-component state with both η_t and η_p and a one-component state with only η_t correspond to the chiral and non-chiral states, respectively. The chiral transition corresponds to a transition from the one-component state to the two-component state due to yielding the second component η_p [31].

2.2 Chiral transition represented with order parameter

Here, we assume that a chiral state for a bulk state below T_{bulk} is identical to that for pure Sr_2RuO_4, as shown in Fig. 2 (b). Considering a point group D_{4h} for a perovskite structure of pure Sr_2RuO_4, some of the possible chiral states are a chiral d-wave, $d_{zx} \pm id_{yz}$, and a chiral p-wave state, $p_x \pm ip_y$, protected by the symmetry of crystal structure, as well as a chiral d-wave, $d_{x^2-y^2} \pm id_{xy}$ in accidental degeneracy [5, 15]. In a traditional classification, $d_{zx} \pm id_{yz}$ and $d_{x^2-y^2} \pm id_{xy}$ are the spin-singlet state, while $p_x \pm ip_y$ is the spin-triplet state [32]. Their intrinsic magnetization is parallel to the z-axis, corresponding to an angular momentum $L_z \neq 0$ for the Cooper pair. The crystalline structure along the z-axis is perpendicular to the RuO_2 layers on the xy-plane, leading to the two-dimensional electron property. Projecting onto the xy-plane of the RuO_2-layer, the components of the order parameter on the xy-plane are common to both chiral states with $d_{zx} \pm id_{yz}$ and $p_x \pm ip_y$ represented using $k_z k_x \pm ik_y k_z$ and $k_x \pm ik_y$, respectively, as shown in Fig. 2 (b) [9, 11, 12, 22, 31].

The chiral states with the time-reversal symmetry breaking are denoted with a combination of two orbital symmetries with a pure imaginary number i, corresponding to $\eta_t + i\eta_p$ in the expression of a superconducting order parameter. Here, the components η_t and η_p of the order parameters correspond to the tangential and perpendicular components of interface Sr_2RuO_2/Ru, respectively. By contrast, a non-chiral state is represented with only one component η_t. Therefore, a chiral transition indicates a transition from the one-component state with η_t to the two-component state with $\eta_t + i\eta_p$, yielding the second component η_p [32].

This interface model sets a superconducting transition temperature that increases near the interface; moreover, it sets boundary conditions for the suppression of the perpendicular components at the interface in the extrapolates of superconductivity to the interface between the Ru-metal and the Sr_2RuO_4-superconductor [9, 11, 12, 22, 31]. The nucleation of superconductivity at the interface originates from the local enhancement of the superconducting transition temperature in a narrow range at the interface on the side of Sr_2RuO_4. In addition, the component η_p, which is perpendicular to the interface, is suppressed by the boundary conditions for the interface. In this situation, the non-chiral state is stabilized with the nucleation of only one component η_t in a zero field at $T_{onset} = 3$ K. This non-chiral state with one component η_t transits to the chiral state with two components $\eta_t + i\eta_p$ at $T^* = 2.3$ K by yielding the second component η_p, owing to the lowering of the temperature in the zero field.

As T^* is a chiral transition temperature due to the lowering of the temperature in the zero field above T^*, the non-chiral state is stabilized with one component η_t in the zero field. When a magnetic field H_z is applied to this non-chiral state with one component η_t above T^*, it transits to the chiral state with two components $\eta_t + i\eta_p$ by yielding the second component η_p as the field-induced chiral transition.

In addition to their chiral state candidates, theoretical studies have also suggested other candidates [4, 15, 26, 28, 29, 34].

2.3 Simulation of field-induced chiral transition

Assuming the chiral states $d_{zx} \pm id_{yz}$ and $p_x \pm ip_y$ as a bulk phase in the eutectic Sr_2RuO_4-Ru, the simulation demonstrates the field-induced chiral phenomena in the inhomogeneous interface phase by applying a z-axis magnetic field H_z parallel to an intrinsic magnetization of the chiral Cooper pair, as shown in Figs. 3, 4, and 5. By numerically solving the Ginzburg–Landau equation, which is set using parameters common to $k_z k_x \pm ik_y k_z$ and $k_x \pm ik_y$ for $d_{zx} \pm id_{yz}$ and $p_x \pm ip_y$, respectively,

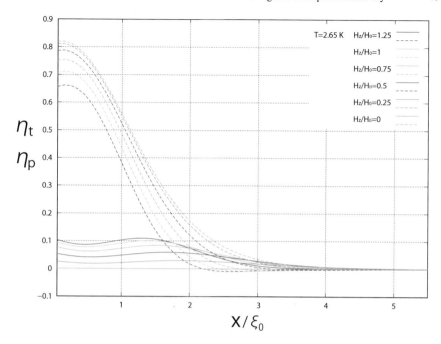

Fig. 3: Two components η_t and η_p of the order parameter dependent on x at $T =$ 2.65 K. The perpendicular component η_p is plotted with solid lines, whereas the tangential component η_t is plotted with dashed lines. An external magnetic field H_z parallel to the z-axis varies in units of H_0, where H_0 is the critical field and ξ_0 is the coherence length of pure Sr_2RuO_4 at $T = 0$. The magnetic field H_z is given in units of $H_0 = 0.075$ T, compared with the experimental critical field of Sr_2RuO_4-Ru [9, 36]. The non-chiral state is stabilized with the nucleation of only one component η_t in a zero field at $T = 2.65$ K, as this temperature is above the chiral transition temperature $T^* = 2.3$ K due to lowering temperature in a zero field. The non-chiral state, i.e., the state with one component η_t, transits to the chiral state, i.e., the state with two components $\eta_t + i\eta_p$, by yielding the second component η_p under the application of the magnetic field H_z, indicating a field-induced chiral transition.

the two components, η_p and η_t, of the superconducting order parameters and the vector potential A are obtained as numerical solutions [9, 11, 12, 22, 31]. The chiral transition is shown with η_p and η_t in Fig. 3, and 4, and paramagnetic and screening supercurrents are calculated from η_p, η_t, and A [9], as shown in Fig. 5.

An interface model sets the critical superconducting temperature enhancing near the interface, as well as the boundary conditions such that a perpendicular component η_p is suppressed at the interface [9, 11, 12, 22, 31]. According to this setting, η_t and η_p depend on the position x from the interface, as shown in Fig.2 (b); thus a superconducting order parameter and a vector potential A depend on the position x. As the intrinsic magnetization and the external field are parallel to the z-axis, and

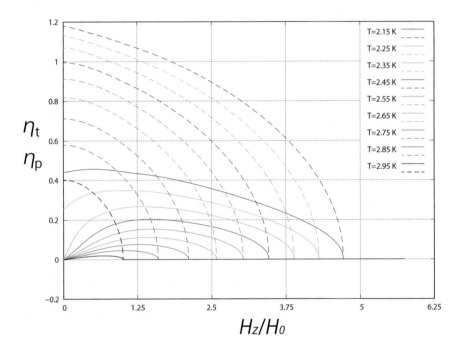

Fig. 4: Dependence of the maximum values of the two components η_t and η_p on an external magnetic field H_z. When the magnetic field is applied to the non-chiral state above $T^* = 2.3$ K, a chiral transition occurs by yielding η_p.

the A includes an intrinsic magnetic field and an external magnetic field H_z parallel to a z-axis, setting $A = (0, A_y, 0)$ connects to a total magnetic field B_z through $B = \nabla \times A$.

In this one-dimensional model, the following is the Ginzburg–Landau equation with the two components, η_t and η_p, of the order parameter for the chiral state [9, 11, 12, 22, 31]

$$a\eta_t + \frac{1}{4}b\eta_t(3\eta_t^2 + \eta_p^2) - K_2\partial_x^2\eta_t + \gamma^2 A_y^2 K_1\eta_t - \gamma K_{3,4}(\partial_x\eta_p A_y + A_y\partial_x\eta_p) = 0,$$

$$a\eta_p + \frac{1}{4}b\eta_p(3\eta_p^2 + \eta_t^2) - K_1\partial_x^2\eta_p + \gamma^2 A_y^2 K_2\eta_p + \gamma K_{3,4}(\partial_x\eta_t A_y + A_y\partial_x\eta_t) = 0,$$

$$(1)$$

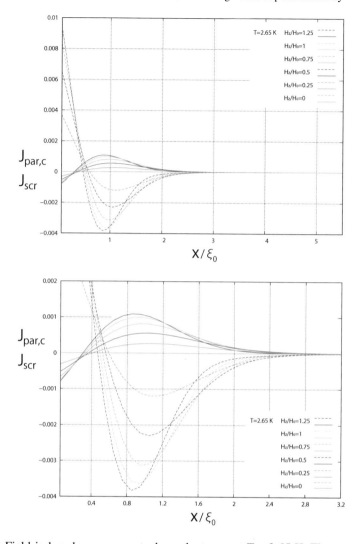

Fig. 5: Field-induced supercurrents dependent on x at $T = 2.65$ K. The upper panel shows the overall view, and the lower panel shows partial magnification. The paramagnetic chiral current $J_{\text{par,c}}$ is plotted with solid lines, and the screening current J_{scr} is plotted with dashed lines. An external magnetic field H_z parallel to the z-axis varies in units of H_0, where H_0 is the critical field and ξ_0 is the coherence length of pure Sr_2RuO_4 at $T = 0$. Both currents have extremely small values, i.e., nearly zero in a zero field, and both currents are induced in a magnetic field. The paramagnetic current $J_{\text{par,c}}$ increases by stabilizing a chiral state by strengthening the magnetic field visible in the order parameter, as shown in Fig. 3, while the screening current also increases by shielding the strengthening magnetic field.

where the parameters for gradient terms are set as $K_2 = K_{3,4} = K_1/3$. The coefficient a is set as $a = a(T) = a'(T - T_c(x))/T_{c,SRO}$ with $a' > 0$ and a x-dependent critical superconducting temperature $T_c(x)$ [9, 11, 12, 22, 31]. An onset of interface superconductivity at 3 K sets with $T_c(x) = T_{c,SRO} + T_0/\cosh(x/w)$ through arranging the narrow width w and the enhancement T_0 [9, 11, 12]. In contrast, the bulk phase is below the superconducting transition temperature $T_{c,SRO}$ of the pure Sr_2RuO_4.

On the other hand, the boundary conditions at the interface at $x = 0$ are set for the interface superconductivity [9] as follows

$$K_1\partial_x\eta_p(x)\big|_{x=0} = \frac{1}{l_p}\eta_p(0) + \gamma A_y(0)K_{3,4}\eta_t(0),$$

$$K_2\partial_x\eta_t(x)\big|_{x=0} = -\gamma A_y(0)K_{3,4}\eta_p(0), \tag{2}$$

where l_p is an extrapolation length in which the superconductivity extrapolates to the Ru-metal through the interface; meanwhile, the perpendicular component η_p is suppressed at the interface. The formulation of the supercurrent [9] is as follows,

$$j_y(x) = 8\pi \left[-\gamma^2 A_y(K_1|\eta_p|^2 + K_2|\eta_t|^2) + \gamma K_{3,4}(\eta_t\partial_x\eta_p - \eta_p\partial_x\eta_t) \right], \tag{3}$$

where the $K_{3,4}$-term is a paramagnetic chiral current, and the K_1, K_2-term is a screening current.

Note that the Ginzburg–Landau equation, i.e., Eq. (1), for $d_{zx} \pm id_{yz}$ and $p_x \pm ip_y$, has a symmetry common to that of $d_{x^2-y^2} \pm id_{xy}$, while the parameters for $d_{zx} \pm id_{yz}$ and $p_x \pm ip_y$ differ from those of $d_{x^2-y^2} \pm id_{xy}$. This common symmetry leads to qualitatively identical field-induced chiral phenomena despite the difference in parameters.

Demonstrating the field-induced chiral transition by applying H_z, Fig. 3 shows the components of the order parameter depending on the distance x from the interface Sr_2RuO_4/Ru. The change in the two components η_t and η_p depends on the magnetic field, which leads to a field-induced chiral transition at $T = 2.65$ K. In a zero field $H = 0$, the state at $T = 2.65$ K is a non-chiral state that stabilizes near T_{onset} as the onset temperature. The non-chiral state represents only one component η_t of the order parameter, i.e., the tangential component, which is enhanced near the interface, whereas the perpendicular component η_p becomes zero by suppression through the effect of the interface. On applying a magnetic field to the non-chiral state in a zero field, the perpendicular component η_p increases from zero and appears as the second component. Subsequently, a transition to a chiral state occurs, representing both the components η_t and η_p, which is the field-induced chiral transition. Here, the components of the order parameter are enhanced near an interface because the inhomogeneous interface state is localized near a Ru-metal inclusion that originates from the enhancement of the superconducting transition temperature near a Ru-metal/Sr_2RuO_4-superconductor interface. In this model, it is assumed that a chiral state exists in the bulk state below T_{bulk}.

Fig. 4 shows the field dependence of the maximum values of the two components η_t and η_p with respect to the distance. The non-chiral state appears with only one component η_t above $T^* = 2.3$ K in a zero field $H = 0$. When an external magnetic field is applied to the non-chiral state, the second component η_p newly appears owing to the field-induced chiral transition. The value of vertical component η_p increases by the strengthening the magnetic field, indicating that the further stabilization of the chiral state. In contrast, the tangential component η_t decreases under the application of the field. Additionally, by further strengthening the field, η_t and η_p decrease, and then both are reduced to zero when the magnetic field is strengthened to the critical magnetic field H_{c2}, where the superconductivity vanishes.

The computational simulation shows the field-induced chiral transition, i.e., a non-chiral state transits to a chiral state under the application of a magnetic field parallel to an intrinsic magnetization in a model that assumes that the bulk phase is in a chiral state. An existing study has reported field-induced chiral stability with the H-T phase diagram in detail [9], which qualitatively consists of the field dependence of a zero-bias anomaly in a differential conductance of quasi-particles, observed via tunneling spectroscopy for the interface of Ru/Sr$_2$RuO$_4$ [14,36]. There is a qualitative agreement between the theoretical and experimental results under the assumption that the bulk state is in a chiral state; this is evidence that a pure Sr$_2$RuO$_4$ has a chiral state [9].

2.4 Paramagnetic chiral supercurrent

Next, we show the paramagnetic chiral supercurrent in Sr$_2$RuO$_4$-Ru in Fig. 5. The field-induced chiral stability causes paramagnetic supercurrents. Moreover, the inversion of chirality occurs at a certain distance. The total supercurrent comprises both the paramagnetic chiral current and the screening current. The paramagnetic current can be attributed to the paramagnetic coupling with an external magnetic field. By contrast, the screening current persists because the superconductivity ejects the external magnetic field [9, 11].

3 Computation of field-induced chiral phenomena

The Ginzburg–Landau equation in Eq. (1) is a variational equation derived from the Ginzburg–Landau free energy. It is a simultaneous differential equation with boundary conditions at the interface. The numerical calculation for solving the equation involves the use of the quasi-Newton method. The flowchart of the calculation process for the algorithm is shown in Fig. 6 (a). As solutions of the simultaneous equation, we obtain two components η_t and η_p of the order parameter and a vector potential A_y, which corresponds to H_z through $\boldsymbol{B} = \boldsymbol{\nabla} \times \boldsymbol{A}$. Consistent solutions are obtained when the iterative calculation process converges, as shown in Fig. 6

(b). The supercurrent is calculated using three solutions: η_t, η_p, and A_y; a paramagnetic chiral current $J_{par,c}$ and a screening current J_{scr} are thus obtained. Meanwhile, the two components η_t and η_p are expressed as changes in the order parameter responsible for the field-induced chiral transition. In addition, the dependence of the order parameter on the distance from the interface indicates an inversion of the intrinsic magnetization as a change in the sign of one component, i.e., η_t [10]. The field-induced chiral transition, paramagnetic chiral current, and inversion of intrinsic magnetization with the distance originate from the paramagnetic coupling of an intrinsic magnetization with an external magnetic field, and the three field-induced phenomena are related by a calculation based on the convergent solutions of the Ginzburg–Landau equation, as shown in Fig. 6 (b).

The long calculation time required for obtaining solutions to the quasi-Newton method depends on the size of the calculation, which varies with the temperature and magnetic field, as well as the mesh number used to divide the distance. We performed this calculation using SX-ACE and the SX-Aurora TSUBASA at the Cybermedia Center, Osaka University, and Cyberscience Center, Tohoku University [2, 16]. The iteration for mesh numbers greater than 240 is performed using vectorization with a vector engine. A long time is required for the numerical calculation when the temperature and magnetic field are sufficiently varied for evaluating the field-induced chiral stability in inhomogeneous superconductivity. Moreover, the required calculation time increases by considering a greater distance to evaluate the dependence of the order parameter on the distance from the interface. In order to reduce this calculation time, code-tuning improves the vectorization rate from 94.4% to 99.4%, significantly reducing the calculation time by 68% [38].

Acceleration via code-tuning makes it possible to analyze the field-induced chiral phenomena in a shorter calculation time. To study the field dependence of the chiral state near the bulk phase, a long distance must be set in the calculation because the order parameter further away from the interface when the temperature is lowered toward T_{bulk}. Moreover, to evaluate the gradient terms of the equation in detail, the mesh number must be increased because the order parameter and vector potential vary depending on the distance. This requires a longer calculation time. Reducing the calculation time is an effective method to increase the distance, and it is possible to extend the simulation to a lower temperature region toward T_{bulk}. At this point, it is valuable to reduce the calculation time through acceleration via optimized code tuning that is suitable for the vector parallelization in the SX-Aurora TSUBASA [38].

4 Summary

As described in this paper, a simulation based on the Ginzburg–Landau equation demonstrated the field-induced chiral phenomena due to paramagnetic coupling of an intrinsic magnetization with an external magnetic field in inhomogeneous superconductivity, such as that in a eutectic system and systems with dislocation or

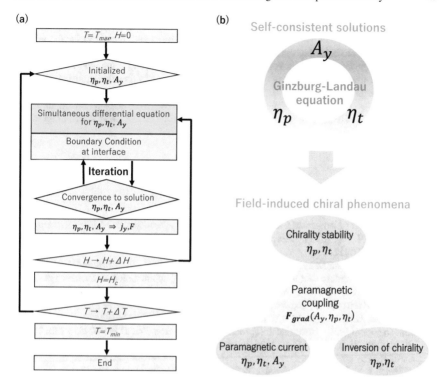

Fig. 6: (a) Field-induced chiral phenomena are the chiral stability, paramagnetic chiral current, and inversion of chirality, accompanied by the gain in the free energy owing to paramagnetic coupling. The simulation based on the Ginzburg–Landau theory shows the relation between the three field-induced phenomena, owing to the chiral response to an external magnetic field. (b) Flowchart of numerical calculation based on the quasi-Newton method.

stress. The simulation demonstrates these phenomena in the interface superconducting model of eutectic Sr_2RuO_4-Ru, and the results of the field-induced chiral transition serve as evidence of the chiral state in the bulk state owing to their good agreement with the experimental results. Using a high-performance computer, the SX-Aurora TSUBASA, for the simulation, the calculation time was reduced through acceleration via code optimization that is suitable for vector parallelization. This reduction of calculation time makes it possible to extend the simulation to a lower-temperature region, which requires the evaluation to be performed considering a longer distance. In addition to Sr_2RuO_4 [19, 21], the simulation can be extended to analyze the inhomogeneous state in other chiral superconductors; field-induced chiral phenomena can also be expected in the inhomogeneous state in other candidates of chiral superconductors, such as UTe_2 with a point group D_{2h} for a crystal structure [8, 27].

Acknowledgements H.K. is grateful to M. Sigrist for the valuable suggestions regarding the theoretical aspects on this work, and acknowledges A. Ramires and Y. Fukaya for the valuable discussions on superconducting states. This work was performed using SQUID at the Cybermedia Center, Osaka University. Further, it was partly performed on the supercomputer SX-ACE at the Cybermedia Center, Osaka University, and Tohoku University. In addition, this study was supported by the Joint Usage and Research of JHPCN (No. jh200032). For the comparison, the performance of the code was evaluated on the supercomputer FUGAKU through the Startup Preparation Project (No. hp200216) of HPCI. This work is also supported by the JSPS Core-to-Core Program No. JPJSCCA20170002.

References

1. J. Bardeen, L.N. Cooper and J.R. Schrieffer. Theory of Superconductivity. *Phys. Rev.* **108**, 1175 (1957). DOI 10.1103/PhysRev.108.1175. URL https://link.aps.org/doi/10.1103/PhysRev.108.1175

2. R. Egawa, K. Komatsu, S. Momose, Y. Isobe, A. Musa, H. Takizawa and H. Kobayashi. Potential of a modern vector supercomputer for practical applications: performance evaluation of SX-ACE. *The Journal of Supercomputing* **73**(9), 3948 (2017). DOI 10.1007/s11227-017-1993-y. URL https://doi.org/10.1007/s11227-017-1993-y

3. S.B. Etter, H. Kaneyasu, M. Ossadnik and M. Sigrist. Limiting mechanism for critical current in topologically frustrated Josephson junctions. *Phys. Rev. B* **90**, 024515 (2014). DOI 10.1103/PhysRevB.90.024515. URL https://link.aps.org/doi/10.1103/PhysRevB.90.024515

4. Y. Fukaya, T. Hashimoto, M. Sato, Y. Tanaka and K. Yada. Spin susceptibility for orbital-singlet Cooper pair in the three-dimensional Sr_2RuO_4 superconductor. *Phys. Rev. Research* **4**, 013135 (2022). DOI 10.1103/PhysRevResearch.4.013135. URL https://link.aps.org/doi/10.1103/PhysRevResearch.4.013135

5. V. Grinenko, D. Das, R. Gupta, B. Zinkl, N. Kikugawa, Y. Maeno, C.W. Hicks, H.H. Klauss, M. Sigrist and R. Khasanov. Unsplit superconducting and time reversal symmetry breaking transitions in Sr2RuO4 under hydrostatic pressure and disorder. *Nature Communications* **12**(1), 3920 (2021). DOI 10.1038/s41467-021-24176-8. URL https://doi.org/10.1038/s41467-021-24176-8

6. V. Grinenko, S. Ghosh, R. Sarkar, J.C. Orain, A. Nikitin, M. Elender, D. Das, Z. Guguchia, F. Brückner, M.E. Barber, J. Park, N. Kikugawa, D.A. Sokolov, J.S. Bobowski, T. Miyoshi, Y. Maeno, A.P. Mackenzie, H. Luetkens, C.W. Hicks and H.H. Klauss. Split superconducting and time-reversal symmetry-breaking transitions in Sr_2RuO_4 under stress. *Nature Physics* **17**, 748 (2021). DOI 10.1038/s41567-021-01182-7. URL https://doi.org/10.1038/s41567-021-01182-7

7. Y. Imai, K. Wakabayashi and M. Sigrist. Effect of the RuO6 Octahedron Rotation at the Sr2RuO4 Surface on Topological Property. *Journal of the Physical Society of Japan* **83**(12), 124712 (2014). DOI 10.7566/JPSJ.83.124712. URL https://doi.org/10.7566/JPSJ.83.124712

8. L. Jiao, S. Howard, S. Ran, Z. Wang, J.O. Rodriguez, M. Sigrist, Z. Wang, N.P. Butch and V. Madhavan. Chiral superconductivity in heavy-fermion metal UTe_2. *Nature* **579**, 523 (2020). DOI 10.1038/s41586-020-2122-2. URL https://doi.org/10.1038/s41586-020-2122-2

9. H. Kaneyasu, Y. Enokida, T. Nomura, Y. Hasegawa, T. Sakai and M. Sigrist. Properties of the $H - T$ phase diagram of the $3 - K$ phase in eutectic Sr_2RuO_4-Ru: Evidence for chiral superconductivity. *Phys. Rev. B* **100**, 214501 (2019). DOI 10.1103/PhysRevB.100.214501. URL https://link.aps.org/doi/10.1103/PhysRevB.100.214501

10. H. Kaneyasu, Y. Enokida, T. Nomura, Y. Hasegawa, T. Sakai and M. Sigrist. Features of Chirality Generated by Paramagnetic Coupling to Magnetic Fields in the 3 K-Phase of Sr_2RuO_4. *JPS Conf. Proc.* **30**, 011039 (2020). DOI 10.7566/JPSCP.30.011039. URL https://journals.jps.jp/doi/10.7566/JPSCP.30.011039

11. H. Kaneyasu, S.B. Etter, T. Sakai and M. Sigrist. Evolution of the filamentary 3-Kelvin phase in Pb − Ru − Sr_2RuO_4 Josephson junctions. *Phys. Rev. B* **92**, 134515 (2015). DOI 10.1103/PhysRevB.92.134515. URL https://link.aps.org/doi/10.1103/PhysRevB.92.134515

12. H. Kaneyasu, N. Hayashi, B. Gut, K. Makoshi and M. Sigrist. Phase Transition in the 3-Kelvin Phase of Eutectic Sr2RuO4–Ru. *Journal of the Physical Society of Japan* **79**, 104705 (2010). DOI 10.1143/JPSJ.79.104705. URL https://doi.org/10.1143/JPSJ.79.104705

13. H. Kaneyasu and M. Sigrist. Nucleation of Vortex State in Ru-Inclusion in Eutectic Ruthenium Oxide Sr2RuO4–Ru. *Journal of the Physical Society of Japan* **79**, 053706 (2010). DOI 10.1143/JPSJ.79.053706. URL https://doi.org/10.1143/JPSJ.79.053706

14. M. Kawamura, H. Yaguchi, N. Kikugawa, Y. Maeno and H. Takayanagi. Tunneling Properties at the Interface between Superconducting Sr2RuO4 and a Ru Microinclusion. *Journal of the Physical Society of Japan* **74**, 531 (2005). DOI 10.1143/JPSJ.74.531. URL https://doi.org/10.1143/JPSJ.74.531

15. S.A. Kivelson, A.C. Yuan, B. Ramshaw and R. Thomale. A proposal for reconciling diverse experiments on the superconducting state in Sr2RuO4. *npj Quantum Materials* **5**(1), 43 (2020). DOI 10.1038/s41535-020-0245-1. URL https://doi.org/10.1038/s41535-020-0245-1

16. K. Komatsu, S. Momose, Y. Isobe, O. Watanabe, A. Musa, M. Yokokawa, T. Aoyama, M. Sato and H. Kobayashi. Performance Evaluation of a Vector Supercomputer SX-Aurora TSUBASA. In: *SC18: International Conference for High Performance Computing, Networking, Storage and Analysis*, pp. 685–696 (2018). DOI 10.1109/SC.2018.00057

17. A.J. Leggett. A theoretical description of the new phases of liquid ^{3}He. *Rev. Mod. Phys.* **47**, 331 (1975). DOI 10.1103/RevModPhys.47.331. URL https://link.aps.org/doi/10.1103/RevModPhys.47.331

18. G.M. Luke, Y. Fudamoto, K.M. Kojima, M.I. Larkin, J. Merrin, B. Nachumi, Y.J. Uemura, Y. Maeno, Z.Q. Mao, Y. Mori, H. Nakamura and M. Sigrist. Time-reversal symmetry-breaking superconductivity in Sr_2RuO_4. *Nature* **394**, 558 (1998). DOI 10.1038/29038. URL https://doi.org/10.1038/29038

19. A.P. Mackenzie, T. Scaffidi, C.W. Hicks and Y. Maeno. Even odder after twenty-three years: the superconducting order parameter puzzle of Sr2RuO4 *npj Quantum Materials* **2**, 40 (2017). DOI 10.1038/s41535-017-0045-4. URL https://doi.org/10.1038/s41535-017-0045-4

20. Y. Maeno, T. Ando, Y. Mori, E. Ohmichi, S. Ikeda, S. NishiZaki and S. Nakatsuji. Enhancement of Superconductivity of Sr_2RuO_4 to 3 K by Embedded Metallic Microdomains. *Phys. Rev. Lett.* **81**, 3765 (1998). DOI 10.1103/PhysRevLett.81.3765. URL https://link.aps.org/doi/10.1103/PhysRevLett.81.3765

21. Y. Maeno, H. Hashimoto, K. Yoshida, S. Nishizaki, T. Fujita, J.G. Bednorz and F. Lichtenberg. Superconductivity in a layered perovskite without copper. *Nature* **372**, 1476 (1994). DOI 10.1038/372532a0. URL https://www.nature.com/articles/372532a0

22. M. Matsumoto, C. Belardinelli and M. Sigrist. Upper Critical Field of the 3 Kelvin Phase in Sr_2RuO_4. *Journal of the Physical Society of Japan* **72**, 1623 (2003). DOI 10.1143/JPSJ.72.1623. URL https://doi.org/10.1143/JPSJ.72.1623

23. M. Matsumoto and M. Sigrist. Quasiparticle States near the Surface and the Domain Wall in a $px \pm ipy$-Wave Superconductor. *Journal of the Physical Society of Japan* **68**, 994 (1999). DOI 10.1143/JPSJ.68.3120. URL https://doi.org/10.1143/JPSJ.68.3120

24. W. Meissner and R. Ochsenfeld. Ein neuer Effekt bei Eintritt der Supraleitfähigkeit. *Naturwissenschaften* **21**, 787 (1933). DOI 10.1007/BF01504252. URL https://doi.org/10.1007/BF01504252

25. H.K. Onnes. Further experiments with liquid helium. D. On the change of electrical resistance of pure metals at very low temperatures, etc. V. The disappearance of the resistance of mercury. *Akad. van Wetenschappen (Amsterdam)* **14**, 113–115 (1911).

26. A. Ramires and M. Sigrist. Superconducting order parameter of Sr_2RuO_4: A microscopic perspective. *Phys. Rev. B* **100**, 104501 (2019). DOI 10.1103/PhysRevB.100.104501. URL https://link.aps.org/doi/10.1103/PhysRevB.100.104501

27. S. Ran, C. Eckberg, Q.P. Ding, Y. Furukawa, T. Metz, S.R. Shanta, I.L. Lin, M. Zic, H. Kim, J. Paglione and N.P. Butch. Nearly ferromagnetic spin-triplet superconductivity. *Science* **365**, 684 (2019). DOI 10.1126/science.aav8645. URL https://www.science.org/doi/abs/10.1126/science.aav8645

28. A.T. Rømer, P.J. Hirschfeld and B.M. Andersen. Superconducting state of Sr_2RuO_4 in the presence of longer-range Coulomb interactions. *Phys. Rev. B* **104**, 064507 (2021). DOI 10.1103/PhysRevB.104.064507. URL https://link.aps.org/doi/10.1103/PhysRevB.104.064507

29. A.T. Rømer, D.D. Scherer, I.M. Eremin, P.J. Hirschfeld and B.M. Andersen. Knight Shift and Leading Superconducting Instability from Spin Fluctuations in Sr_2RuO_4. *Phys. Rev. Lett.* **123**, 247001 (2019). DOI 10.1103/PhysRevLett.123.247001. URL https://link.aps.org/doi/10.1103/PhysRevLett.123.247001

30. T. Shiroka, R. Fittipaldi, M. Cuoco, R. De Renzi, Y. Maeno, R.J. Lycett, S. Ramos, E.M. Forgan, C. Baines, A. Rost, V. Granata and A. Vecchione. μSR studies of superconductivity in eutectically grown mixed ruthenates. *Phys. Rev. B* **85**, 134527 (2012). DOI 10.1103/PhysRevB.85.134527. URL https://link.aps.org/doi/10.1103/PhysRevB.85.134527

31. M. Sigrist and H. Monien. Phenomenological Theory of the 3 Kelvin Phase in Sr_2RuO_4. *Journal of the Physical Society of Japan* **70**, 2409 (2001). DOI 10.1143/JPSJ.70.2409. URL https://doi.org/10.1143/JPSJ.70.2409

32. M. Sigrist and K. Ueda. Phenomenological theory of unconventional superconductivity. *Rev. Mod. Phys.* **63**, 239 (1991). DOI 10.1103/RevModPhys.63.239. URL https://link.aps.org/doi/10.1103/RevModPhys.63.239

33. A. Steppke, L. Zhao, M.E. Barber, T. Scaffidi, F. Jerzembeck, H. Rosner, A.S. Gibbs, Y. Maeno, S.H. Simon, A.P. Mackenzie and C.W. Hicks. Strong peak in T_c of Sr_2RuO_4 under uniaxial pressure. *Science* **355**(6321), eaaf9398 (2017). DOI 10.1126/science.aaf9398. URL https://www.science.org/doi/abs/10.1126/science.aaf9398

34. H.G. Suh, H. Menke, P.M.R. Brydon, C. Timm, A. Ramires and D.F. Agterberg. Stabilizing even-parity chiral superconductivity in Sr_2RuO_4. *Phys. Rev. Research* **2**, 032023 (2020). DOI 10.1103/PhysRevResearch.2.032023. URL https://link.aps.org/doi/10.1103/PhysRevResearch.2.032023

35. J. Xia, Y. Maeno, P.T. Beyersdorf, M.M. Fejer and A. Kapitulnik. High Resolution Polar Kerr Effect Measurements of Sr_2RuO_4: Evidence for Broken Time-Reversal Symmetry in the Superconducting State. *Phys. Rev. Lett.* **97**, 167002 (2006). DOI 10.1103/PhysRevLett.97.167002. URL https://link.aps.org/doi/10.1103/PhysRevLett.97.167002

36. H. Yaguchi, K. Takizawa, M. Kawamura, N. Kikugawa, Y. Maeno, T. Meno, T. Akazaki, K. Semba and H. Takayanagi. Spectroscopy of Sr_2RuO_4/Ru Junctions in Eutectic. *AIP Conference Proceedings* **850**, 543 (2006). DOI 10.1063/1.2354825. URL https://aip.scitation.org/doi/abs/10.1063/1.2354825

37. Y. Ying, N. Staley, Y. Xin, K. Sun, X. Cai, D. Fobes, T.J. Liu, Z.Q. Mao and Y. Liu. Enhanced spin-triplet superconductivity near dislocations in Sr_2Ru_O. *Nat. Com.* **4**, 2596 (2013). DOI 10.1038/ncomms3596. URL https://www.nature.com/articles/ncomms3596

38. S. Yoshida, A. Endo, H. Kaneyasu and S. Date. First Experience of Accelerating a Field-Induced Chiral Transition Simulation Using the SX-Aurora TSUBASA. *Supercomput. Front. and Innov.* **8**, 43 (2021). DOI 10.14529/jsfi210203. URL https://superfri.org/index.php/superfri/article/view/383

Exploiting Hybrid Parallelism in the LBM Implementation Musubi on Hawk

Harald Klimach, Kannan Masilamani and Sabine Roller

Abstract In this contribution we look into the efficiency and scalability of our Lattice Boltzmann implementation *Musubi* when using *OpenMP* threads within an *MPI* parallel computation on *Hawk*. The Lattice Boltzmann method enables explicit computation of incompressible flows and the mesh discretization can be automatically generated, even for complex geometries. The basic Lattice Boltzmann kernel is fairly simple and involves only few floating point operations for each lattice node. A simple loop over all lattice nodes in each partition of the *MPI* parallel setup lends to a straight forward loop parallelization with *OpenMP*. With increased core counts per compute node, the use of threads on the shared memory nodes is gaining importance, as it avoids overly small partitions with many outbound communications to neighboring partitions. We briefly discuss the hybrid parallelization of *Musubi* and investigate how the usage of *OpenMP* threads affects the performance when running simulations on the *Hawk* supercomputer at *HLRS*.

1 The Lattice Boltzmann method

The Lattice Boltzmann method (*LBM*)[9] offers an efficient explicit method to compute incompressible or weakly compressible flows by modelling the gas with a discrete velocity space for the Boltzmann equation in a mesoscopic scale. For the discretization a regular mesh is used, usually with cubic cells, and the connections to the neighbors offer the discrete velocity directions to be considered in the numerical method. Hence, *LBM* can be considered as a stencil method, with similar properties in communication and parallelization as other mesh-based methods. An advantage of *LBM* can be observed in the treatment of boundaries. Due to the use of discrete

Harald Klimach, Kannan Masilamani and Sabine Roller
DLR e.V., Institut für Softwaremethoden zur Produkt-Virtualisierung, Zwickauer Str. 45, 01069 Dresden,
e-mail: harald.klimach@dlr.de, kannan.masilamani@dlr.de, sabine.roller@dlr.de

velocities, boundaries only have to be considered along those discrete directions and accurate boundaries can be obtained by intersecting the one dimensional lines in cubical boundary cells with the surfaces describing geometrical boundaries. Such line intersections with surfaces can be computed robustly and are, thus, well suited for automated mesh generations.

The most commonly used stencil in three dimensions are *D3Q19* and *D3Q27*. The *D3Q19* makes use of 18 neighbors together with the state at rest resulting in 19 values of the probability density function to describe the fluid state. In this stencil, 18 neighbors are all immediately connected cells except for those at the corners of the cube (6 sides and 12 edges of the cube). Other stencil *D3Q27* is required in some *LBM* approaches which makes use of all 26 immediate neighbors.

The lattice Boltzmann equation with classical collision operator from Bhatnagar, Gross and Krook *BGK*[1] is given by

$$f_i(\mathbf{x} + \mathbf{c}_i \Delta t, t + \Delta t) - f_i(\mathbf{x}, t) = \Omega_i(\mathbf{x}, t) \tag{1}$$

where $f_i(\mathbf{x}, t)$ is the probability density function at the position vector $\mathbf{x}$ and at time step t along the discrete direction i; Δt is the discrete time step; $\mathbf{c}_i$ is the discrete velocities and Ω_i is the collision operator. There is a multitude of collision operations available. However, here we will only consider the classical operation described by Bhatnagar, Gross and Krook (*BGK*) as

$$\Omega_i = -\frac{1}{\tau}(f_i(\mathbf{x}, t) - f_i^{eq}(\mathbf{x}, t)) \tag{2}$$

where $\mathbf{f_i^{eq}}$ is the Maxwell-Boltzmann distribution function. For weakly-compressible flows, it is given by

$$f_i^{eq}(\rho, \mathbf{u}) = \omega_i \rho \left(1 + \frac{(\mathbf{c}_i \cdot \mathbf{u})}{c_s^2} + \frac{(\mathbf{c}_i \cdot \mathbf{u})^2}{2c_s^4} - \frac{(\mathbf{u} \cdot \mathbf{u})}{2c_s^2}\right), \tag{3}$$

where ω_i are the lattice weights and $c_s = c/\sqrt{3}$ is the speed of sound in lattice. $c = \Delta x/\Delta t$ is the lattice velocity where Δx is the discretization size. ρ and $\mathbf{u}$ are macroscopic density and velocities which are computed from probability density function by

$$\rho = \sum_{i=1}^{Q} f_i \tag{4}$$

and

$$\rho \mathbf{u} = \sum_{i=1}^{Q} \mathbf{c}_i f_i. \tag{5}$$

The pressure P is calculated from the density ρ using the equation of state relation as $P = c_s^2 \rho$. The relaxation time τ in Eq. 2 is related to kinematic viscosity as

$$v = c_s^2 \Delta t (\tau - 0.5). \tag{6}$$

In LBM, Eq. 1 is solved in two steps: streaming and collision. Streaming is exchanging the probability density functions of the particles along their respective directions with neighboring cells and collision is computing a new state in each cell according to collision operator. In general, the collision requires only few floating point operations per cell.

2 The Musubi implementation

Musubi[3] is our open source implementation of the Lattice Boltzmann method, mostly written in Fortran 2003 and primarily parallelized with via the Message Passing Interface (*MPI*) [6]. It makes use of an octree mesh discretization with cubical cells. Cells in the mesh are sorted according to the Morton or Z curve [5], which provides some maintaining of the multi dimensional locality in the one dimensional sorted list of cells. The solver works on the cells according to that ordering and the mesh partitioning is achieved by splitting the ordered list of cells into equally sized chunks. A double buffer is used to hold the state and allow the access to the previous iteration in the streaming step. Some additional arrays are used to hold further auxiliary values for all cells. All in all we find a computational intensity of around $1/3$ floating point operations per Byte.

Meshes can be created with the mesh generator *Seeder* [2], which provides the mesh in this form of an ordered list of cells or for simple meshes. This allows for a distributed reading of the mesh information as each process can identify the part of the file it needs to read with little information on the mesh. Alternatively, simple meshes like the ones we will consider here, can also be generated by *Musubi* itself. The list of cells may be sparse and thus, explicit neighborhood information is needed to address the stencil cells. Hence, the stencil implementation here behaves as an unstructured mesh with indirect addressing of the stencil cells. However, the known topology of the octree and the ordering according to the space-filling curve enables the identification of neighbor cells across partitions in the distributed memory parallel computation. Therefore, nearly arbitrary stencils can be employed and *Musubi* makes use of that in the implementation of the various *LBM* schemes. Meshes might have cells on different levels of the octree refinement, but on each level the kernel just acts as if working on uniform mesh. This is achieved by ghost cells that provide interpolated values from other refinement levels. Due to this behavior it is possible to perform some assessment of fundamental properties of this kind of kernel in a single level uniform mesh, which we look at in this contribution.

2.1 OpenMP in Musubi

The *OpenMP* parallelization[8] in Musubi is incomplete and various features do not yet benefit from it. But the parallelization of the fundamental kernel is straight forward, as it essentially is a single loop over all cells to update the lattice nodes. An *OpenMP* parallel region is put around these loops to realize the shared memory parallelism within *MPI* processes. The *MPI* communication itself is not put into a parallel region and does not profit from shared memory parallelization. With a static schedule the loop parallelism this way results in a partitioning similar to the *MPI* partitioning as the cells are sorted according to the space-filling curve. Accordingly the expectation is that the degree of parallelism can be shifted interchangeably between the one and the other.

3 Hybrid parallelization

In supercomputing systems a hierarchy of parallelism and data access can be observed. The most obvious decomposition can be observed in the construction of large clusters from individual nodes. Where individual nodes provide shared memory access between all processing units within it. The number of processing units within such a node mostly depends on the number of cores we find in the employed processors. And accordingly, we observe a growing degree of shared memory parallelism within those nodes as the number of cores in modern processors increases. Using a distributed memory parallelization concept uniformly for all processing units is possible, but results in small partitions that end up with many individual neighbor partitions that may be located on other nodes. This results in a larger number of smaller network communications between nodes. A strategy to minimize this effect and obtain larger *MPI* partitions with fewer, but larger network communications, is to resemble this two-level hierarchy of the hardware in the application. With *OpenMP* in each *MPI* partition processing units can be dedicated to parallel work in a reduced number of distributed memory partitions. Such a strategy than results in a less fragmented communication pattern across the network of the cluster.

Unfortunately, there are also some downsides involved in the hybrid parallelization. The management of threads results in some overheads and we increase the risk encounter resource conflicts in commonly used resources in the node, like shared caches or memory interfaces. As long as there are parts of the code that do not benefit from *OpenMP* parallelization, we face the problem that some parts do not benefit from a larger number of threads, but would benefit from more *MPI* processes. In the end the optimal choice depends at least on the system, the application is run on. It also depends on the specific setup to be run, but here we want to look at properties of the fundamental kernel, which may also be instructive in a wider context.

4 Performance assessment on Hawk

To evaluate the effect of shifting parallelism between *MPI* and *OpenMP* on *Hawk*, we run *Musubi* for various problem sizes on a range of node counts. These runs are performed with different numbers of threads per process, such that always all cores are participating in the computation.

4.1 The Hawk computing system

The *Hawk* computing system installed at the High Performance Computing Center Stuttgart (*HLRS*) is based on the *AMD EPYC 7742* processor[7], which has 64 cores operating at 2.25 GHz and *AVX2* vector instructions, yielding a theoretical peak performance of 2.3 TFLOPS. Each node has two of these processors and, thus, has a total of 128 physical cores. There are groups of 4 cores that share their L3 cache and build a so called *CoreCompleX*. Two of those *CoreCompleXs* are paired together and share one of the 8 memory channels in the processor. Finally two of those memory channels are put into a NUMA node per socket. Accordingly we see 16 physical cores in each NUMA node and with the two sockets in each computing node a total of 8 NUMA nodes. We will consider an *OpenMP* parallelism of up to 16 threads per *MPI* process, which corresponds to one *MPI* process per NUMA node. A further increase in the degree of shared memory parallelism is expected to incur degrading performance in comparison to a distributed memory strategy, due to the strong hierarchical structure of the memory access paths. Because of the shared L3 cache, a natural choice for the degree of shared memory parallelism in this system is 4 threads per process, putting all cores in a *CoreCompleX* to work on a shared memory region. Each memory channel provides a bandwidth of around 24 GB/s resulting in a total of 192 GB/s per socket or 384 GB/s per node to access the capacity of 256 GB.

4.2 The Musubi setup

For this contribution we use *Musubi* in version *9ccba4387413* [4]. It is using the environment offered by the modules *gcc/9.2.0* and *mpt/2.23* and with *OpenMP* support. The simulation setup is a small initial pressure pulse in a cubic domain of edge length 10 that is periodic in all directions. This simple mesh can be generated by *Musubi* during the simulation and can be easily scaled up in factors of 8. The initial spherical pulse is located in the center of the domain, has an amplitude of 1.2 over a background value of 1 and a halfwidth of 1. As collision operator we use *BGK* and we look at the *D3Q19* and *D3Q27* stencils. Each simulation is executed for at least

5 minutes of running time. The executable is run with the following command, with *nprocs* representing the number of *MPI* processes and *tpp* the number of *OpenMP* threads per process:

```
mpirun -np $nprocs omplace -tm pthreads -nt $tpp
```

4.3 Results

Performance for Lattice Boltzmann methods is usually measured in million lattice updates per second *MLUPS* and we consider this measure here per node. This measure is independent of the actual running time and allows for a comparison between different runs and simulations.

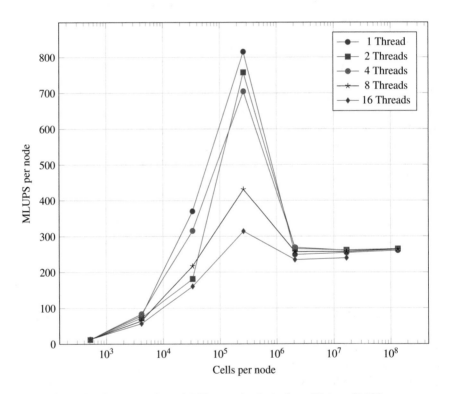

Fig. 1: Performance for *D3Q19* on a single node, utilizing all 128 cores.

As described above we perform runs with varying problem sizes on a node and record the resulting performance measure in *MLUPS*. Figure 1 shows the resulting graph for the *D3Q19* stencil on a single node. This form of representation nicely shows the variation of the performance with the problem size. Overheads, like communication, dominate for very small problem sizes at the left end of the graph. Then a peak can be observed where the overall problem is still sufficiently small to completely fit into the caches, avoiding slower memory accesses. Finally, a relatively flat performance plateau is reached for larger problem sizes, until it does not fit into the memory of the node anymore.

What we can observe in this single node analysis is that in the region with memory access with more than a million cells, there is basically no performance difference between 1, 2, 4 and 8 *OpenMP* threads per *MPI* process. With 16 threads the performance is clearly reduced, which seems to indicate that it is important for *Musubi* that an *MPI* process is not spread across multiple memory channels. In the cache region we also see a clear diminishing of the performance for 8 *OpenMP* threads, where the shared memory of a process spans across two *CoreCompleXs* and accordingly two shared L3 caches.

For the *D3Q27* stencil a similar behavior can be seen in figure 2. With 27 discrete velocity directions to represent the state, more memory is required to represent the state in each cell and less cells fit into the memory of the single node than with 19 directions only. The largest domain that still fit into memory for the *D3Q19* stencil $(134, 217, 728$ cells), therefore, does not fit here anymore and the graphs, and the largest mesh we compute is $16, 777, 216$ cells large. Note, that this is a lot smaller than what would fit into the memory, which would be more than 90 million cells.

This single node analysis shows the principal behavior of the *LBM* implementation in *Musubi* on each node of *Hawk* and illustrates the performance impact of the different parallelization strategies on the hierarchical memory layout of the system. We can note that the use of 4 *OpenMP* threads on as many cores yields roughly the same performance as a *MPI*-only parallelization and in the region with memory access up to 8 threads can be used interchangeably to *MPI* parallelism.

The analysis on a single node, however, does not show how the use of *OpenMP* threads influences the network communication between nodes. As stated above a motivation to make use of *OpenMP* parallelism is to reduce the number of individual communication partners with whom comparably small messages need to be exchanged. To assess this, we repeat these runs on larger node counts with 8, 64 and 512 nodes. Incrementing by a factor of 8 yields here the same problem sizes per node again, and allows for a direct comparison of the individual data points.

For brevity we only depict the corresponding graphs for 512 nodes. In this case we have $65, 536$ cores working in parallel. This analysis is shown in figure 3 for *D3Q19* and in figure 4 for *D3Q27*. As can be seen in these figures, the behavior on 512 nodes is quite similar to the one on a single node. However, we also observe some differences. Most importantly we now see that for small problem sizes per node a higher performance is achieved with 2 and 4 threads and not with the pure *MPI* parallel computation.

In confirmation to the observation for a single node it appears reasonable to make use of a single *MPI* process for each *CoreCompleX* with the cores that share their L3 cache also sharing their memory address space. Shared memory parallelization beyond that diminishes the performance in the cache region with small cell counts per node, but for larger problems with the need to access the memory, also larger

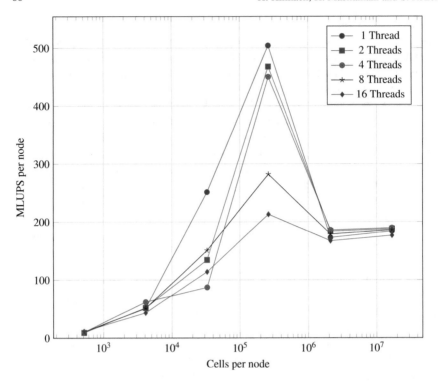

Fig. 2: Performance for *D3Q27* on a single node, utilizing all 128 cores.

shared memory processes with up to 8 cores (sharing one memory channel) can be utilized. Though for smaller problems, where all elements would fit into the caches, a larger performance decrease can be observed for those shared memory partitions spanning more than a single *CoreCompleX*.

Using 16 cores, spanning two memory channels in a NUMA node, as a shared memory parallel group within an *MPI* incurs too many drawbacks in the memory access of the hierarchical processor design to be used efficiently by *Musubi* also on 512 nodes.

For the scaling we look at the *D3Q27* stencil as the more memory and communication intensive scheme and stick to the parallelization with one *MPI* process per *CoreCompleX* and 4 *OpenMP* threads per process to allow concurrent computation on the 4 physical cores. As we have seen in the above measurements this configuration nicely fits the physical properties of the processor and provides good performance across problem sizes.

We also include the computation on 2048 nodes here, though these do not result in exactly the same number of cells per node as the other runs. This is the maximal number of nodes available to users in the regular queue on *Hawk* and provides 262, 144 physical cores for parallel execution. The resulting performance per node is illustrated in figure 5.

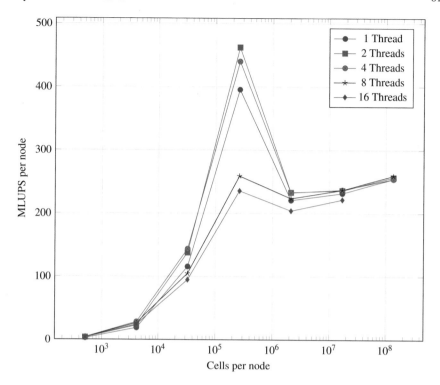

Fig. 3: Performance per node for *D3Q19* on 512 nodes, utilizing all 65,536 cores.

This illustration shows that there is a significant performance degradation per node in the region of small problem size that fit into the cache from 450 MLUPS on a single node to 293 MLUPS per node on 2048 nodes. Due to the fast computation without accessing memory outside the caches, the necessary communication on the larger node counts increasingly dominates the execution time in this region. Nevertheless, there is still a higher performance observed in this cache region than when accessing the memory for larger problems per node. Without *OpenMP* parallelism the performance drops further down to 255 MLUPS per node. In the region with memory access, however, the performance degradation is relatively small dropping from 190 MLUPS on a single node to 166 MLUPS per node on 512 nodes for 16,777,216 cells per node.

As observed above, the *OpenMP* parallelism does not have much of an influence for problem sizes per node and for other numbers of threads a similar behavior is observed. And the performance for 16,777,216 cells per node on 512 nodes does not vary much with the number of threads per process. This is summarized in table 1.

Of a little more interest in this respect is the strong scaling, where the problem size per node decreases with growing numbers of nodes. Figure 7 shows the strong scaling efficiency for the various number of threads per process. Aside from the rapid

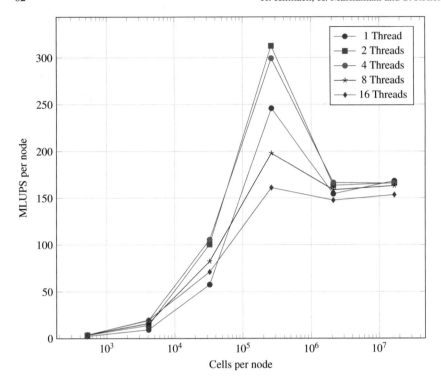

Fig. 4: Performance per node for *D3Q27* on 512 nodes, utilizing all 65, 536 cores.

Table 1: Performance per node on 512 nodes for 16, 777, 216 cells per node

Threads per process	MLUPS per node
1	168
2	166
4	166
8	163
16	153

decline in the parallel efficiency beyond the peak in the cache region we see that the use of *OpenMP* threads here allows for a better scaling to small problems per node, with 4 threads per process, matching the *CoreCompleX* yielding the highest parallel efficiency on 512 nodes. Note, that this graph is somewhat truncated due to the few cells fitted on a single node, though more than 6 times as many could fit into the memory.

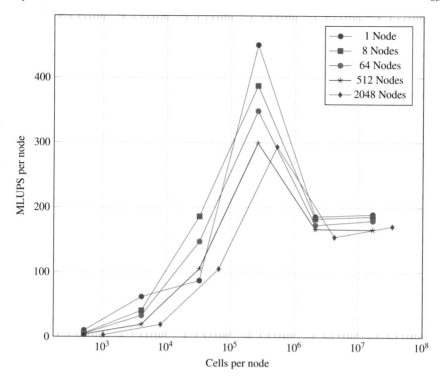

Fig. 5: Performance per node for *D3Q27* with 4 *OpenMP* threads per *MPI* process.

5 Conclusion

We have presented a basic analysis of the performance behavior of *Musubi* on the *HLRS* computing system *Hawk*. It reveals that up to 4 *OpenMP* threads per process can be used interchangeably with *MPI* parallelism and can slightly improve the performance in strong scaling for very small problems per node. This number of threads corresponds to the *CoreCompleX* of the *AMD EPYC 7742* processors, which groups 4 physical cores that share a L3 cache together. The largest problem computed in this analysis contained $68,719,476,736$ cells and was computed on $262,144$ cores.

Acknowledgements We thank the High-Performance Computing Center Stuttgart (*HLRS*) for the computing time on *Hawk* to perform the presented analysis.

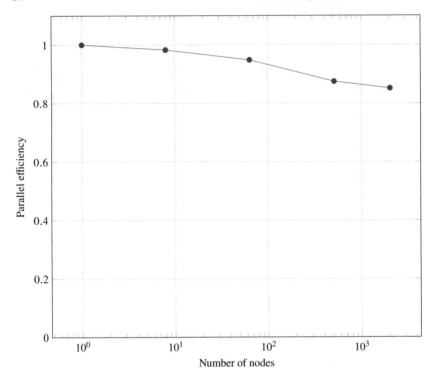

Fig. 6: Weak scaling parallel efficiency for *D3Q27* with 4 *OpenMP* threads per *MPI* process and 16,777,216 cells per node (linear interpolated for 2048 nodes.

References

1. P.L. Bhatnagar, E.P. Gross and M. Krook. A Model for Collision Processes in Gases. I. Small Amplitude Processes in Charged and Neutral One-Component Systems. *Phys. Rev.* **94**(3), 511–525, (1954).
2. D. Harlacher, M. Hasert, H. Klimach, S. Zimny and S. Roller. Tree Based Voxelization of STL Data. In: *High Performance Computing on Vector Systems 2011*, M. Resch, X. Wang, W. Bez, E. Focht, H. Kobayashi and S. Roller, pp. 81–92, Springer Berlin (2012).
3. M. Hasert, K. Masilamani, S. Zimny, H. Klimach, J. Qi, J. Bernsdorf and S. Roller. Complex fluid simulations with the parallel tree-based Lattice Boltzmann solver Musubi. *J. Comp. Sci.* **5**(5), 784–794, (2014).
4. H. Klimach. Musubi Mercurial Repository. https://osdn.net/projects/apes/scm/hg/musubi/. Accessed 2022-03-25.
5. G.M. Morton. A computer oriented geodetic data base and a new technique in flie sequencing. Technical report, IBM Ltd. (1966).
6. MPI: A Message Passing Interface 4.0. https://www.mpi-forum.org/docs/mpi-4.0/mpi40-report.pdf. Accessed 2022-03-25.
7. NASA: HECC Knowlegebase: AMD Rome Processors. https://www.nas.nasa.gov/hecc/support/kb/amd-rome-processors_658.html. Accessed 2022-03-25.

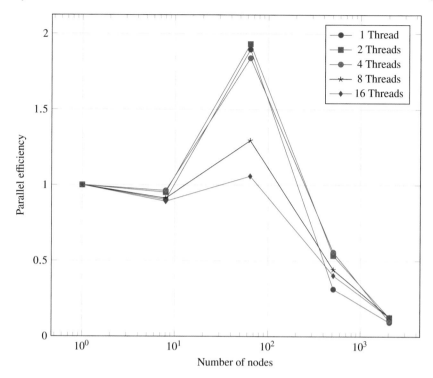

Fig. 7: Strong scaling parallel efficiency for *D3Q27* and 16,777,216 cells in total.

8. OpenMP Application Programming Interface 5.2.
https://www.openmp.org/wp-content/uploads/OpenMP-API-Specification-5-2.pdf. Accessed 2022-03-25.
9. S. Succi. *The Lattice Boltzmann Equation for Fluid Dynamics and Beyond.* Oxford Univ. Press (2001).

MPI Continuations And How To Invoke Them

Joseph Schuchart and George Bosilca

Abstract Asynchronous programming models (APM) are gaining more and more traction, allowing applications to expose the available concurrency to a runtime system tasked with coordinating the execution. While MPI has long provided support for multi-threaded communication and non-blocking operations, it falls short of adequately supporting the asynchrony of separate but dependent parts of an application coupled by the start and completion of a communication operation. Correctly and efficiently handling MPI communication in different APM models is still a challenge. We have previously proposed an extension to the MPI standard providing operation completion notifications using callbacks, so-called MPI Continuations. This interface is flexible enough to accommodate a wide range of different APMs. In this paper, we discuss different variations of the callback signature and how to best pass data from the code starting the communication operation to the code reacting to its completion. We establish three requirements (efficiency, usability, safety) and evaluate different variations against them. Finally, we find that the current choice is not the best design in terms of both efficiency and safety and propose a simpler, possibly more efficient and safe interface. We also show how the transfer of information into the continuation callback can be largely automated using C++ lambda captures.

1 Background

The Message Passing Interface (MPI) offers a host of nonblocking operations, which are started in a procedure call that immediately returns and provides a request handle representing the operation [4]. At the time of this writing, the only way to know whether an operation has completed is to poll for its completion, either by periodically testing the request or by blocking until its completion in a waiting procedure call. This

Joseph Schuchart and George Bosilca
Innovative Computing Laboratory (ICL), University of Tennessee Knoxville (UTK),
1122 Volunteer Blvd, Knoxville, TN 37996, U.S.A., e-mail: schuchart@icl.utk.edu

M. M. Resch et al. (eds.), *Sustained Simulation Performance 2021*,
https://doi.org/10.1007/978-3-031-18046-0_5

poses a significant challenge for applications utilizing asynchronous programming models such as OpenMP [9] or higher level distributed runtime systems managing communication through MPI because the requests have to be stored and (repeatedly) passed back into MPI to determine their status.

Over time, several approaches have been proposed that try to hide the synchronizing incurred by waiting for an MPI operation to complete, including TAMPI [6] and the integration of lightweight threads into MPI libraries [3]. These approaches attempt to block and switch the execution context until operations have completed. However, all such approaches are dependent on the support for specific threading implementations and thus not portable.

MPI Continuations, on the other hand, have been proposed as a way to minimize the request management overhead in applications or runtime systems by attaching a callback to a single or a set of continuations [7]. The callback will be executed once all of the operations the continuation was attached to have completed. The application or runtime system can then react to that change in state inside the callback, e.g., by enqueuing a new task or releasing resources associated with that operation. This approach has shown promising results in both OpenMP task-based applications as well as when integrated with the PaRSEC runtime system [8].

A similar approach, dubbed MPI Detach, has been proposed concurrently [5]. While conceptually similar to the MPI Continuations proposal, the callback interface proposed passes a status (or an array of statuses) into the continuations, which would require additional memory management by the MPI library.

In this work, we explore the design of the callback signature of MPI Continuations, focusing on usability, potential performance pitfalls, and safety concerns stemming from the necessary memory management. The rest of this paper is structured as follows: Section 2 provides a short overview over the current state of the continuations proposal. Section 3 discusses various requirements we impose on the design of the callback signature. Section 4 discusses various variations of the callback interface together with their benefits and drawbacks. Section 5 demonstrates the use of the continuations interface in the context of C++. Section 6 draws our conclusions from this exploration.

2 Current state

The Continuations proposal introduces two new concepts into MPI: Continuations and Continuation Requests (CRs). Continuations are a tuple of a callback function and a state on which the callback function operates. Similar concepts can be found in other instances employing the concept of continuations, e.g., continuations proposed for C++ futures in the form of $std::future<T>::then()$ [1], which accepts a callable object (e.g., a lambda with it's capture context) that takes the value of type T of the future as its sole parameter. Here, the code in the lambda's body is the callback function while its captured context and the value of type T are the state to operate on. The HPX and UPC++ programming systems relies heavily on continuations on C++ futures [2, 10].

2.1 Continuations

MPI Continuations are created using either MPI_Continue or MPI_Continueall which will **attach** a continuation to a single request or a set of requests, respectively. Since MPI currently only provides C and Fortran interfaces, automatic C++-style context captures cannot be directly supported. Thus, a user-provided data pointer is accepted that will be passed to the continuation callback. This data pointer represents the context of the continuation and is never dereferenced by MPI. It is thus of little relevance to the discussion in this work.

However, an operation in MPI is represented by a request and further information about the outcome of the operation can be gathered from status objects obtained for each request upon its completion (e.g., the tag and sender process rank in the case of a receive operation, or an error code in case of faults). In the case of MPI_Wait, a request is passed together with a pointer to a status object. The status object is optional and the application may pass MPI_STATUS_IGNORE instead, in which case no further information about the operation will be made available. In its current form, a pointer to a single status object or an array of status objects may be passed to MPI_Continue and MPI_Continueall, respectively, and the status objects will be set before the continuation is invoked. This pointer will then also be passed as an argument to the continuation.

2.2 Continuation Requests (CR)

Continuation Requests serve a dual purpose. First, they provide an abstract handle to a set of continuations **registered** with this CR, allowing the application to poll for the completion of all registered continuations and (by extension) the associated operations. Once all registered continuations have completed, a wait or test procedure call on that CR will signal its completion (by returning from wait or setting flag = 1 in a test). Second, CRs provide a facility for progressing outstanding communication operations and to execute eligible continuations.

The relation between CRs, continuations, and operations is shown in Figure 1: multiple continuations may be registered with one continuation request but each continuation may only be registered with a single CR. The latter is a consequence of the fact that continuations are not accessible explicitly through a handle and their lifetime is managed entirely by MPI. Similarly, a continuation may be attached to multiple MPI operations at once, causing the callback to be executed once all of the are complete. However, each MPI operation may only be associated with one continuation. The transitive closure of these relations is that a CR represents one or many MPI operations and that a successful test on a CR implies the completion of all MPI operations associated with continuations registered with that CR.

Fig. 1: Relations between Continuation Requests (CR), Continuations, and MPI Operations: multiple continuations can be registered with the same CR (left) and a Continuation can be attached to multiple operations (right). However, only continuation can be attached to any given MPI Operation.

2.3 Current API design

Listing 1 shows the current API as proposed. The ownership of non-persistent requests is returned to MPI and the respective entry in the array is set to MPI_REQUEST_NULL. The ownership of persistent requests is not changed. This behavior is similar to that of an optional array of status(es) (or MPI_STATUS[ES]_IGNORE otherwise) is passed to the function. The statuses will be set to the statuses of the completed MPI operations before the continuation callback is invoked and the pointer to the statuses provided by the user is passed as the first argument.

As a second argument, the user_data pointer is passed to the callback. This pointer may reference any state the continuation may require for its execution.

In addition to requests, statuses, the callback function pointer, and the user-provided state, the two functions listed in Listing 1 also accept a set of OR-combined flags that control different aspects of the continuation. Among these flags is MPI_CONT_IMMEDIATE to control whether the continuation may be executed immediately if all operations have completed already. If that flag is not set, the continuation will be enqueued for later execution, e.g., when waiting on the continuation request. However, the details of these flags are still fluid and beyond the scope of this paper and not relevant for the ensuing discussion.

As a last argument, the *continuation request* described in Section 2.2 is passed to the attaching functions.

Figure 2 shows the flow of ownership in the current API design. The call to MPI_Isend allocates a request object and passes its ownership back to the caller (who *borrows* it), who is then responsible for releasing that request in a call to MPI_Test or MPI_Wait. If the request is passed to MPI_Continue, its ownership is transferred back to the MPI library, who is then responsible for releasing the associated internal resources. If a status argument other than MPI_STATUS_IGNORE is provided, the ownership of the status buffer is transferred to MPI and the application should not modify the buffer before the continuation is invoked, which implies the

```
typedef void (MPI_Continue_cb_function)(MPI_Status *statuses,
                                        void *user_data);
```

(a) Callback signature.

```
int MPI_Continue(                          int MPI_Continueall(
  MPI_Request *op_req,                       int count,
  MPI_Continue_cb_funtion *cb,               MPI_Request op_req[],
  void *user_data,                           MPI_Continue_cb_funtion *cb,
  int flags,                                 void *user_data,
  MPI_Status *status,                        int flags,
  MPI_Request cont_req);                     MPI_Status statuses[],
                                             MPI_Request cont_req);
```

(b) Attaching to single operation.   (c) Attaching to multiple operations.

Listing 1: API for attaching a continuation to a single or multiple MPI operations.

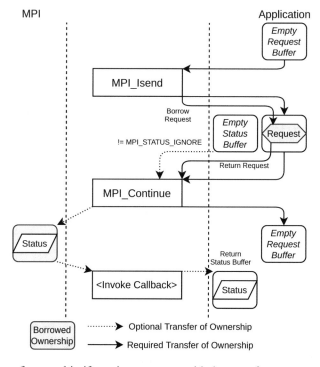

Fig. 2: Flow of ownership if passing a user-provided array of statuses to the continuation.

transfer of ownership of that buffer back to the application. While transient in nature, ownership of the request buffer is transferred into `MPI_Isend` and implicitly returned at the end of the call. We have included these transient ownership transfers for the sake of completeness.

We note that if `MPI_STATUS_IGNORE` is provided instead of a status buffer the only object(s) whose ownership is transferred are the requests. In that case, no borrowed ownership remain after the call to `MPI_Continue`.

3 Callback interface requirements

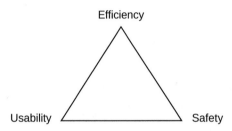

Fig. 3: Usability, efficiency, and safety are often detrimental in the design of APIs, requiring carefully balancing of these three requirements.

We outline three main requirements for the continuations API that we believe should be fundamental to the design of the continuations API. As shown in Figure 3, requirements for a safe, efficient, and easily usable API are often detrimental and need careful balancing.

3.1 Efficiency

The complexity of polling for the completion of requests using existing mechanisms such as `MPI_Testall` and `MPI_Waitall` involves checking the status of each request and progressing communication if required. MPI implementations have been carefully optimized to avoid dynamic memory management in such critical execution paths.

The cost of attaching a continuation to a set of requests and managing its execution should be equally low. In particular, requiring memory allocations that are not strictly necessary and copying objects (e.g., requests and statuses) should be avoided wherever possible. Ideally, no dynamic memory management would be required on the part of the application, at least in the simplest of use-cases. Similarly, requiring the allocation of buffers inside MPI to hold requests or statuses in the design of the API would negatively impact performance even for simple cases.

3.2 Safety

While APIs for the C language rarely can eliminate all possible mistakes made by programmers, good API design aims at minimizing complexities and reducing the probability of such mistakes. In the context of asynchronous execution APIs such as continuations, likely sources of errors are accessing memory in the callback that points to the stack of the function that started the operation and attached the continuation, e.g., trying to access the request or status objects. Ideally, the MPI Continuations API helps users avoid the pitfalls of memory lifetime issues by eliminating disambiguities about object lifetime and ownership.

3.3 Usability

While a clean interface with little or no potential pitfalls certainly contributes to the usability of an interface, some simplifications in the API may require additional steps to achieve complex setups, e.g., management of additional memory (with a potential impact on performance) or set up of custom data structures and the resulting additional code that has to be written and maintained. On the other hand, a complex callback design providing a rich set of information (request handles, status objects, datatypes, message element counts) directly to the callback function may reduce the work on the part of the application since all relevant information is provided directly. However, most application may not need the provided information in their callbacks, resulting in overhead in memory space and time that does not yield any benefits for these applications.

4 Callback interface variations

We will discuss a set of variations in the design of the current API described in Section 2.3, using a simple example

Using the current API, Listing 2 provides an example of attaching a continuation to a nonblocking receive operation. All the continuation does is to enqueue a task that will process the message and release the buffer. The buffer is not processed directly in order to keep the duration of the callback as short as possible and to potentially defer the processing of the message to another thread. There is no use of the status provided when attaching the continuation and the message buffer is passed directly

```
1   /* Continuation request, initialized elsewhere */
2   MPI_Request cont_req;
3
4   void complete_cb(MPI_Status *status, void *buffer) {
5     enqueue_processing_task(buffer);
6   }
7
8   void start_receive(void *buffer, int from, int size){
9     MPI_Request op_req;
10    MPI_Irecv(buffer, size, MPI_BYTE, from, /*tag=*/101,
11              MPI_COMM_WORLD, &op_req);
12    MPI_Continue(&op_req, &complete_cb, buffer, 0,
13                 MPI_STATUS_IGNORE, cont_req);
14  }
```

Listing 2: Simple example of a continuation attached to a nonblocking receive.

```
1   /* Continuation request, initialized elsewhere */
2   MPI_Request cont_req;
3
4   void complete_cb(MPI_Status *status, void *buffer) {
5     int msg_size;
6     MPI_Get_count(status,MPI_BYTE, &msg_size);
7     enqueue_processing_task(buffer, msg_size);
8     free(status);
9   }
10
11  void start_receive(void *buffer, int from, int buffer_size){
12    MPI_Request op_req;
13    MPI_Status *status = malloc(sizeof(MPI_Status));
14    MPI_Irecv(buffer, buffer_size, MPI_BYTE, from, /*tag=*/101,
15              MPI_COMM_WORLD, &op_req);
16    MPI_Continue(&op_req, &complete_cb, buffer, 0,
17                 status, cont_req);
18  }
```

Listing 3: Simple example of a continuation attached to a nonblocking receive operation, querying the status of the operation.

on to the continuation. The value provided for the `status` parameter of the callback will be `MPI_STATUS_IGNORE`. No dynamic has to be allocated in this example. We note that the `cont_req` used in this and the following examples would have been initialized at an earlier point.

Listing 3 provides a variation of this example where `start_receive` posts a receive for a message with a maximum size and uses the status of the operation to query the size of the message actually received. The status is allocated on the heap (using `malloc` in Line 13) to ensure that the memory remains valid until the continuation has executed. The allocated status is subsequently freed in Line 8.

A more complex example employing a persistent receive operation is provided in Listing 4. When attaching the continuation, a status is passed that will be set before the callback is invoked. Like before, the status is allocated on the heap. Instead of

```
1   /* Wrapper around data needed in the callback */
2   typedef struct callback_data_t {
3     MPI_Request op_req; /* persistent operation request */
4     void *msg;            /* message buffer */
5   } callback_data_t;
6
7   void complete_cb(MPI_Status *status, void *user_data) {
8     int cancelled;
9     int msg_size;
10    MPI_Test_cancelled(status, &cancelled);
11    if (cancelled) { /* nothing to be done */
12      free(user_data);
13      free(status);
14      return;
15    }
16    MPI_Get_count(status,MPI_BYTE, &msg_size);
17    /* copy the message and restart the receive */
18    callback_data_t* cb_data = (callback_data_t*)user_data;
19    copy_msg_and_enqueue_task(cb_data->msg, msg_size);
20    MPI_Start(&op_req->op_req);
21    MPI_Continue(&op_req->op_req, &complete_cb,
22                 user_data, 0, status, cont_req);
23  }
24
25  MPI_Request
26  start_recurring_receive(void *buffer, int from, int size) {
27    /* Allocate the callback data */
28    callback_data_t *cbdata = malloc(sizeof(callback_data_t));
29    /* Allocate the status object */
30    MPI_Status *status = malloc(sizeof(MPI_Status));
31    cbdata->msg = buffer;
32    MPI_Recv_init(buffer, size, MPI_BYTE, from, /*tag=*/101,
33                  MPI_COMM_WORLD, &cbdata->op_req);
34    /* start the operation and attach continuation */
35    MPI_Start(&cbdata->op_req);
36    MPI_Continue(&op_req->op_req, &complete_cb,
37                 cbdata, 0, status, cont_req);
38    return op_req->op_req;
39  }
40
41  void stop_recurring_receive(MPI_Request op_req) {
42    MPI_Cancel(&op_req);
43    MPI_Request_free(&op_req);
44  }
```

Listing 4: A more complex example attaching a continuation to a persistent receive. An incoming message will be copied and a task processing it enqueued. The persistent receive is then restarted before the continuation is attached anew. Eventually, the persistent receive will be canceled, which will be detected inside the continuation in Lines 8–14.

just passing the message pointer, this time a structure of type `callback_data_t` is allocated that wraps the pointer to the message and the persistent operation request, both of which are accessed inside the continuation callback. In contrast to C++ lambda captures, such capturing has to be done manually in C.

The start-attach cycle is broken once the persistent receive is cancelled (Lines 41–44) and the check of the status in Line 10 detects the cancellation. The heap memory is released and no task is enqueued to process the message (Lines 12 – 13).

It is not hard to see that the current variant is neither fool-proof nor the most efficient solution. A subtle change to the way the status is allocated can lead to disastrous consequences. If instead of allocating the status on the heap the status was allocated on the stack (as is commonly the case when calling `MPI_Test`), the memory pointed to by `status` in the callback is invalid as it points to the stack of `start_recurring_receive` that is no longer active. Changing the code of Listing 4 accordingly, yields

```
MPI_Status status;
...
MPI_Continue(&op_req->op_req, &complete_cb,
             cbdata, 0, &status, cont_req);
```

Unfortunately, this rather subtle bug is not easy to spot and in practice would likely slip through a code review. It is not unnatural for users to believe that the `status` argument of the callback points to a status object provided by MPI, instead of simply being the status pointer provided while attaching the continuation. Thus, this is a potential source of grave errors that (as all bugs related to memory management) would be hard to debug.

In terms of efficiency, it is questionable why the status should be allocated separately. It would indeed be more efficient to allocate the status as part of the `callback_data_t` structure. However, since a pointer to that structure is already passed to the callback, passing a separate pointer to the callback function seems superfluous. In essence, the status pointer has to be stored and passed twice. Consequently, the current interface breaks with some of the requirements laid out in Section 3, both potentially impairing safety and efficiency.

4.1 Passing requests and user data

Instead of passing the pointer to the status object, it might be tempting to provide request (or array of requests) to the callback and query their status using `MPI_Request_get_status`. After all, unlike the status argument the (array of) request(s) is a non-optional parameter to `MPI_Continue` and `MPI_Continueall`. This would remove the status from the continuations interface altogether and avoid the potential access out-of-scope stack memory from within the continuation callback.

In principle, two sub-variants of this approach are possible.

4.1.1 MPI-provided Request Buffer

The first sub-variant is to allocate an MPI-internal buffer and copy the request or requests passed to `MPI_Continue` or `MPI_Continueall` into it. The ownership of non-persistent requests would still be transferred back to MPI and their handle be replaced by `MPI_REQUEST_NULL` in order to make them inaccessible outside of the continuation callback. This buffer of copied request handles would then be passed into the continuation callback and (together with all non-persistent requests) destroyed after the continuation completes. The flow of ownership is depicted in Figure 4.

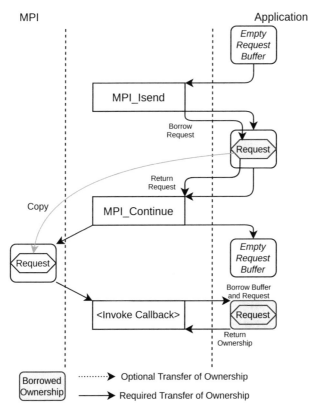

Fig. 4: Flow of ownership if passing an array of copied request handles to the continuation.

A significant drawback of this approach is the required copying of requests and additional dynamic memory management inside the MPI library since the number of requests to which a continuation is attached is not known *a priori*.

4.1.2 User-provided Request Buffer

Instead of allocating a buffer inside the MPI implementation, the API could also directly pass on the pointer to the request(s) provided to `MPI_Continue` or `MPI_Continueall`. As stated in Section 2.3, the ownership of non-persistent requests is returned to MPI. In this case, if the application wanted to access the status of an operation, it would have to request that the request handles are retained. This could be accomplished by introducing and passing a flag such as `MPI_CONT_RETAIN`, requesting that even non-persistent requests are retained until the continuation is invoked. The flow of ownership in this case is depicted in Figure 5.

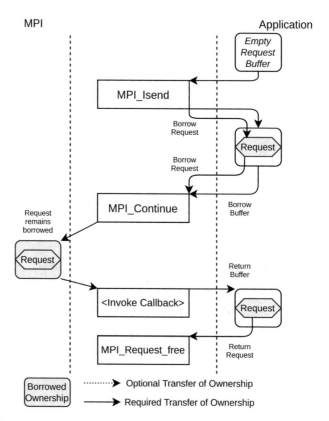

Fig. 5: Flow of ownership if passing the user-provided array of requests to the continuation.

Since ownership would remain with the application, it is necessary to either implicitly (at the end of the continuation) or explicitly return ownership at the end of the callback by invoking `MPI_Request_free` on each non-persistent request. In the interest of efficiency (and symmetry with test and wait functions), the addition of `MPI_Request_freeall` should be considered in this case. For continuations that do not inquire the status of the operation, another flag should be introduced that prevents the implementation from storing and passing on the pointer to the request. This is especially important for requests that were located on the stack, as is the case in Listings 2 and 3. However, a new handle will have to be introduced to pass as an invalid pointer to a request handle.[1]

It becomes apparent that passing the request instead of the status into the continuation breaks with the requirements outlined in Section 3. Either the MPI library is required to allocate internal memory for each continuation, or the issue of pointers potentially pointing to invalid (stack) memory is shifted from the status object to the request objects. On top of that, the added complexity of properly managing the lifetime of requests through flags and additional release of requests opens the door for additional errors in user code and potentially impairs usability.

4.2 Passing only user data

To avoid the potential mistakes in the lifetime management of statuses and requests and the potential efficiency issues outlined in the previous sections, the state passed to the continuation should be reduced to a single pointer. This removes any ambiguity regarding the ownership of the status and request objects and avoids any additional memory allocations.

The code of Listing 4 adapted to only passing the user pointer is provided in Listing 5. It should be noted that while this interface removes potential issues around memory management (and thus provides improved safety and efficiency) a slightly higher burden is put on users in that all state of the continuation has to be collected in a single structure. However, we believe that this is a cost that is worth paying in exchange for the reduced potential of memory management mistakes and efficiency issues.

In order to further reduce the risk of using out-of-scope stack variables in continuations (e.g., from users allocating `callback_data_t` on the stack), the MPI Continuations interface would again have to copy the contents of the user-provided buffer into an internal buffer and pass that buffer to the callback. However, as stated earlier, this may compromise efficiency and safety and does not guarantee that nested pointers do not point to variables on the out-of-scope stack.

With this interface, the simple code in Listing 2 will remain the same, except that the `MPI_Status*` argument to the callback disappears. No dynamic memory allocation would be required in this case. The code in Listing 3 will have to allocate a structure containing the pointer to the message buffer and the status. The allocation is thus shifted from the status object to the user-provided data pointer.

---

[1] MPI does not typically employ the `NULL` pointer but instead defines special values for all invalid handles.

```
1    typedef struct callback_data_t {
2      MPI_Request op_req; /* persistent operation request */
3      MPI_Status status; /* status of the operation */
4      void *msg; /* the message to be received */
5    } callback_data_t;
6
7    void complete_cb(void *user_data) {
8      callback_data_t* cbdata = (callback_data_t*)user_data;
9      int cancelled;
10     MPI_Test_cancelled(&cbdata->status, &cancelled);
11     if (cancelled) { /* nothing to be done */
12       free(cbdata);
13       return;
14     }
15     enqueue_process_process(cbdata->msg);
16     MPI_Start(&op_req->op_req);
17     MPI_Continue(&op_req->op_req, &complete_cb,
18                  cbdata, 0, status, cont_req);
19   }
20
21   MPI_Request
22   start_recurring_receive(void *buffer, int from, int size) {
23     callback_data_t *cbdata = malloc(sizeof(callback_data_t));
24     cbdata->msg = buffer;
25     MPI_Recv_init(buffer, size, MPI_BYTE, from, /*tag=*/101,
26                   MPI_COMM_WORLD, &cbdata->op_req);
27     MPI_Start(&cbdata->op_req);
28     MPI_Continue(&op_req->op_req, &complete_cb,
29                  cbdata, 0, &cbdata->status, cont_req);
30     return cbdata->op_req;
31   }
```

Listing 5: The example of Listing 4 passing the user data pointer as the only state to the continuation callback.

5 C++ lambda capture

A variation of the code of Listing 5 using C++ lambda captures is listed in Listing 6. In this case, the compiler captures all data necessary inside the lambda defined in Lines 30 – 35. Unfortunately, the `status` of the operation cannot be automatically captured by the lambda because it's values is known only once the callback is invoked. Thus, the status is encapsulated inside a wrapper `cb_t` that makes it accessible both inside and outside the lambda.

All other variables are captured *by value* (including the operation request) and stored as part of the `fn` member of `cb_t`. The lambda is marked as `mutable` because both `MPI_Start` and `MPI_Continue` take a non-const pointer to it. The static `invoke` member function of `cb_t` (Lines 10 – 19) will be called by MPI, which then checks for cancellation and invokes the lambda, passing a reference to itself, allowing the lambda to reattach the continuation using the same object.

```
1   /* Callback wrapper typed on the callable's type */
2   template<typename Fn>
3   struct cb_t {
4     /* Status must be accessible outside the callable */
5     MPI_Status status;
6     Fn fn;
7     cb_t(Fn&& fn) : fn(std::forward<Fn>(fn)) {}
8     /* static function invoked from MPI,
9        dispatching to the provided callable */
10    static void invoke(void*data) {
11      cb_t* cb = static_cast<cb_t*>(data);
12      int cancelled;
13      MPI_Test_cancelled(&cb.status, &cancelled);
14      if (cancelled) {
15        delete cb; /* cleanup the wrapper */
16        return;
17      }
18      cb->fn(*cb);
19    }
20  };
21
22  MPI_Request
23  start_recurring_receive(void *buffer, int from, int size) {
24    MPI_Request op_req;
25    MPI_Recv_init(buffer, size, MPI_BYTE, from, /*tag=*/101,
26                  MPI_COMM_WORLD, &op_req);
27    MPI_Start(&op_req);
28    auto cb = new cb_t(
29      /* Marked mutable to pass op_req as non-const. */
30      [=](auto& cb) mutable {
31        process(buffer);
32        MPI_Start(&op_req);
33        MPI_Continue(&op_req, &cb.invoke,
34                     &cb, 0, &cb.status, cont_req);
35      });
36    MPI_Continue(&op_req, &cb->invoke,
37                 cb, 0, &cb->status, cont_req);
38    return op_req;
39  }
```

Listing 6: The example of Listing 4 using C++ lambda capture. The `status` must be accessible inside and outside the lambda expression and thus cannot be captured. Instead it is held in a wrapper object through which the lambda is invoked.

We have deliberately avoided the use `std::function` in order to provide the compiler with the opportunity to inline the code in the lambda, further reducing the overhead of the call. While the use of `std::function` would remove the template parameter Fn from `cb_t`, it introduces a second indirect call in the continuation (in addition to the indirect call to `invoke` inside the MPI library).

We note that the templating of `cb_t` makes it easily composable and reusable, allowing it to be used with different lambdas throughout the an application. While not entirely void of complexity, the use of C++ lambda captures with MPI Continuations removes the hassle of manually transferring data from the callsite into the callback through custom structures, as is required when using C.

6 Conclusions

In this paper, we have discussed several variants of the MPI Continuations API and how state relevant to the execution of the continuation callback can be captured and passed to the callback. We set out a three requirements (usability, efficiency, and safety) that at times are at odds with each other. We found that the interface currently proposed encourages inefficient (by separately allocating the status object(s)) and potentially unsafe (by passing stack-based variables) code. We believe that an interface that requires the aggregation of all variables accessed inside the continuation callback into a single structure yields a safer and potentially more efficient API design. We have also shown that by employing modern C++ lambda captures, this task can be mostly automated. Based on our findings, we will adapt the interface in our MPI Continuations proposal to only pass a single state pointer and to avoid potential confusions about lifetime and ownership of status objects present in the current proposal.

References

1. N. Gustafsson, A. Laksberg, H. Sutter and S. Mithani. N3857: Improvements to `std::future<T>` and Related APIs. Tech. Rep. N3857 (2014).
2. H. Kaiser, T. Heller, B. Adelstein-Lelbach, A. Serio and D. Fey. HPX: A Task Based Programming Model in a Global Address Space. In: *Proceedings of the 8th International Conference on Partitioned Global Address Space Programming 'Models*, PGAS '14, pp. 6:1–6:11, ACM (2014). DOI 10.1145/2676870.2676883
3. H. Lu, S. Seo and P. Balaji. MPI+ULT: Overlapping Communication and Computation with User-Level Threads. In: *2015 IEEE 17th International Conference on High Performance Computing and Communications, 2015 IEEE 7th International Symposium on Cyberspace Safety and Security, and 2015 IEEE 12th International Conference on Embedded Software and Systems* (2015). DOI 10.1109/HPCC-CSS-ICESS.2015.82
4. MPI: A Message-Passing Interface Standard, Version 4.0. Tech. rep. (2021). URL https://www.mpi-forum.org/docs/mpi-4.0/mpi40-report.pdf
5. J. Protze, M.A. Hermanns, A. Demiralp, M.S. Müller and T. Kuhlen. MPI Detach – Asynchronous Local Completion. In: *27th European MPI Users' Group Meeting*, EuroMPI/USA '20, Association for Computing Machinery (2020). DOI 10.1145/3416315.3416323
6. K. Sala, X. Teruel, J.M. Perez, A.J. Peña, V. Beltran and J. Labarta. Integrating blocking and non-blocking MPI primitives with task-based programming models. *Parallel Computing* **85**, 153–166 (2019). DOI 10.1016/j.parco.2018.12.008

7. J. Schuchart, C. Niethammer and J. Gracia. Fibers are not (P)Threads: The Case for Loose Coupling of Asynchronous Programming Models and MPI Through Continuations. In: *27th European MPI Users' Group Meeting* EuroMPI/USA '20, pp. 39–50, Association for Computing Machinery (2020). DOI 10.1145/3416315.3416320
8. J. Schuchart, P. Samfass, C. Niethammer, J. Gracia and G. Bosilca. *Parallel Computing* **106**, 102793 (2021). https://doi.org/10.1016/j.parco.2021.102793. URL https://www.sciencedirect.com/science/article/pii/S0167819121000466
9. J. Schuchart, K. Tsugane, J. Gracia and M. Sato. The Impact of Taskyield on the Design of Tasks Communicating Through MPI. In: *Evolving OpenMP for Evolving Architectures* (Springer International Publishing, 2018), pp. 3–17. DOI 10.1007/978-3-319-98521-3_1. Awarded Best Paper
10. Y. Zheng, A. Kamil, M.B. Driscoll, H. Shan and K. Yelick. UPC++: A PGAS Extension for C++. In: *2014 IEEE 28th International Parallel and Distributed Processing Symposium*, pp. 1105–1114 (2014). DOI 10.1109/IPDPS.2014.115

Xevolver for Performance Tuning of C Programs

Hiroyuki Takizawa, Shunpei Sugawara, Yoichi Shimomura, Keichi Takahashi and
Ryusuke Egawa

Abstract We introduce a C interface for standard C programmers to define their own
code transformation rules for performance tuning, mainly assuming loop transformations. The proposed C interface can support most of important features provided
by the Fortran interface. As a result, performance concerns can be defined separately
as user-defined code transformation rules, and thus the original application code can
be kept unchanged as much as possible.

1 Introduction

High-Performance Computing (HPC) applications are often specialized for their
target platforms to achieve reasonably high performance. Such code specialization
is not only labor-intensive, but also makes it difficult to migrate the applications to
other platforms. One idea to overcome this difficulty is separation of performance
concerns, meaning that the information specific to a particular platform is expressed
separately from the computation. However, in reality, one bad practice heavily used
in HPC application development to achieve high performance on multiple platforms
is a so-called "ifdef" approach that writes multiple code versions within a single file
and uses C macro conditionals for the preprocessor to switch the code versions to

Hiroyuki Takizawa, Yoichi Shimomura and Keichi Takahashi
Cyberscience Center, Tohoku University,
e-mail: takizawa@tohoku.ac.jp, shimomura32@tohoku.ac.jp, keichi@tohoku.ac.jp

Shunpei Sugawara
Graduate School of Information Sciences, Tohoku University,
e-mail: shunpei@hpc.is.tohoku.ac.jp

Ryusuke Egawa
Tokyo Denki University, e-mail: egawa@mail.dendai.ac.jp

© The Author(s), under exclusive license to Springer Nature Switzerland AG 2023   85
M. M. Resch et al. (eds.), *Sustained Simulation Performance 2021*,
https://doi.org/10.1007/978-3-031-18046-0_6

be used at the compilation, severely degrading the code maintainability. Therefore, we need an effective way of expressing platform-specific performance concerns separately from application codes.

In the Xevolver project [16], we have developed a programming framework for performance tuning with user-defined code transformations [6, 12]. A high-level programming interface for standard HPC programmers to describe their own code transformation rules has also been developed mainly for Fortran codes [10], because most of legacy HPC applications are written in Fortran. Lately, however, it is gradually becoming popular to use not only Fortran but also other programming languages such as C and C++, especially when new HPC applications are developed from scratch. Moreover, there are many tools such as CIVL [17] available for C and C++, but not for Fortran. If we need to use such a tool, there is no choice to use Fortran at the HPC application development. Therefore, we consider that Xevolver should provide a high-level interface not only for Fortran but also for C.

In this article, we introduce a C interface for standard C programmers to define their own code transformation rules for performance tuning, mainly assuming loop transformations. Through various case studies [5, 13, 15], Xevolver's approach has been proven to be effective in achieving high performance and code maintainability. The proposed C interface can support most of important features provided by the Fortran interface. As a result, performance concerns can be defined separately as user-defined code transformation rules, and thus the original application code can be kept unchanged as much as possible.

2 Related work

Software automatic performance tuning, or auto-tuning (AT) for short, is indispensable to exploit the performance of modern HPC systems by empirically exploring a parameter space relevant to performance [7]. To use AT techniques, an application code must be developed to be *auto-tunable* [14], and be able to change its behaviors according to parameter tuning and code version switching. One challenging issue is that there is no established way of developing a practical application while keeping it auto-tunable.

So far, several case studies have demonstrated that Xevolver's approach can enable standard HPC programmers to define their own code transformation rules without any special knowledge about compiler implementation technologies [5, 13, 15]. Although an HPC application code is directly modified by hand to adapt to its target platform in many cases, such manual code modifications can be replaced with code transformations, and thus the original HPC application code can remain almost unchanged if Xevolver can translate the original code to its optimized and/or auto-tunable version right before the compilation process.

Although the original Xevolver framework [12] has only low-level interfaces to manipulate internal code representation, Xevtgen [10] has been developed to provide a high-level interface for Fortran programmers to describe code transformation rules using Fortran syntax. Thanks to the high-level interface, Xevolver enables to develop a Fortran code without specializing it for any specific platform.

Egawa et al. [4] have presented a database of performance tuning expertise, called HPC refactoring catalog. Loop optimization techniques in the database are described along with Fortran sample codes, and the loop optimization is expressed as a code transformation rule. Sugawara et al. [11] demonstrated that most of those techniques are also effective for C programs running on recent platforms.

3 Xevolver for C

We are now designing and developing Xevolver for C (Xev-C) for performance tuning of C programs using user-defined code transformations.

In the original Xevolver framework [12], Abstract Syntax Trees (ASTs) are expressed in an XML format, called XML-AST. Then, AST-based code transformation rules are internally expressed also in another XML format, called XSLT, which is a standardized format to describe transformations of XML data, and hence can be used for transformation of XML-AST data. Since it is too painful for standard HPC programmers to describe XSLT rules to define code transformation, Xevtgen has been developed to define code transformation rules using Fortran syntax familiar to HPC programmers [10]. AST-based representation of code transformation is certainly useful for Xevolver to express a wide variety of code transformations. However, our case studies show that the high-level interface provided by Xevtgen can cover most cases where Xevolver's approach is required. Moreover, in the case where AST-based transformation is appropriate, there are many other tools to express such a code transformation. Therefore, Xev-C internally uses Clang AST, and implements only the high-level interface for HPC programmers to define their own code transformation rules required in practice on a case-by-case basis. Xev-C does not explicitly expose ASTs to users, and assumes to use Clang tools to develop AST-based code transformations if necessary.

Unlike the original Xevolver framework built on top of the ROSE compiler infrastructure [8], Xev-C is implemented using Clang [1]. Xev-C takes two C files for user-defined code transformations as shown in Figure 1. One of the two C files is an application code to be transformed, and the other is a code transformation rule written in C. As with Xevtgen, Xev-C assumes that users provide two versions of a code fragment to define a code transformation. One is the original version, and the other is its transformed version. Figure 2 shows an example of code transformation rule defined using Xev-C. If the rule in Figure 2 is applied to the code in Figure 3, the first loop is exactly the same as the original code version in the rule, and thus is

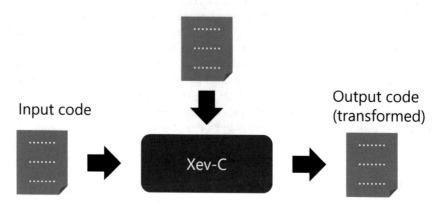

Fig. 1: Overview of Xevolver for C (Xev-C). Xev-C takes two C codes, input code and transformation rule, and then produces one code, a transformed version of the input code. All the codes are written in C.

```
1    #include "xev_defs.h"
2
3    int i,j;
4    double a[10][10], b[10], c[10];
5
6    int main()
7    {
8      xev_stmt_src("label1");
9      {
10       for(i=0;i<10;i++){
11         for(j=0;j<10;j++){
12           c[i] += a[i][j]*b[j];
13         }
14       }
15     }
16
17     xev_stmt_dst("label1");
18     {
19       for(int k=0;k<100;k++){
20         i = k/10;
21         j = k%10;
22         c[i] += a[i][j]*b[j];
23       }
24     }
25   }
```

Fig. 2: A simple transformation rule for loop collapse.

```
1   #include <stdio.h>
2
3   int i,j,n;
4   double a[10][10], b[10], c[10], d[10];
5
6   int main()
7   {
8     read_data_from_file(a,b,c,d);
9
10    for(i=0;i<10;i++){
11      for(j=0;j<10;j++){
12        c[i] += a[i][j]*b[j];
13      }
14    }
15
16    for(i=0;i<10;i++){
17      for(n=0;n<10;n++){
18        d[i] += a[i][n]*c[n];
19      }
20    }
21
22    write_data_to_file(a,b,c,d);
23
24    return 0;
25  }
```

Fig. 3: A simple code to be transformed.

replaced with the transformed version. As a result, in this particular example, loop collapse is applied to the first loop in Figure 3, but not to the second loop whose loop index of the innermost loop is n.

The transformation rule in Figure 2 is just text replacement and transforms a loop only if the loop is exactly identical to the original version in the rule. A large number of rules would be required if performance tuning is done only with such text replacement rules. To achieve performance tuning with as few rules as possible, Xev-C provides special variables, called Xev variables, so that a rule can be defined not for a particular code fragment but for a code pattern. Figure 4 shows a rule of loop collapse similar to the rule in Figure 2. In Figure 4, Xev variables xi, xj, and stmt are defined and used in the rule. Since xi and xj represent any expressions, they match any variables. Even if the loop index has a different name, the rule can be applied to the loop. Similarly, since stmt represents any statements, statements in the loop body do not affect to determine if the rule is applied. If the rule is applied, the loop body is unchanged and simply copied to the transformed version. As a result, in Figure 3, the second loop as well as the first one will be transformed by the rule.

```
1   #include "xev_defs.h"
2
3   int i,j;
4   xev_expr xi,xj;
5   xev_stmt* stmt;
6
7   int main()
8   {
9     xev_stmt_src("label1");
10    {
11      for(xi=0;xi<10;xi++){
12        for(xj=0;xj<10;xj++){
13          stmt;
14        }
15      }
16    }
17
18    xev_stmt_dst("label1");
19    {
20      for(int k=0;k<100;k++){
21        xi = k/10;
22        xj = k%10;
23        stmt;
24      }
25    }
26  }
```

Fig. 4: A simple transformation rule with Xev variables for loop collapse.

4 Evaluation and discussions

In this work, we have examined that the performance tuning expertise recorded in HPC refactoring catalog [4] can be expressed using the current design of Xev-C. As discussed in [11], Fortran codes in 28 out of 31 cases in the catalog can be translated into C. The three cases not translated into C use either of using built-in Fortran functions or libraries available only in Fortran. Most of the performance tuning techniques in the catalog are vectorization-aware loop optimizations mainly targeting the previous-generation vector systems, SX-9 [9] and SX-ACE [3]. In the following evaluation, the performance gains by the techniques are evaluated on the latest vector systems, two generations of SX-Aurora TSUBASA [2]. The system specifications are summarized in Table 1.

Xev-C does not support all the features provided by Xevtgen yet. However, we have confirmed that all the code transformation rules in the 28 C codes can be expressed as Xev-C rules. Therefore, we believe that the expressive ability of the current design of Xev-C is high enough at least for vectorization-aware loop optimizations.

Figure 5 shows the performance gains by the code transformations for SX-Aurora TSUBASA. The vertical axis shows the speedup ratio of the transformed code to the original code for each system. Each code is complied with either of -02 or -04, to discuss how compiler optimization affects the performance. Overall, most of vectorization-aware loop optimizations for the previous-generation systems

Table 1: Specifications of the two generations of SX-Aurora TSUBASA used in the evaluation.

2nd generation SX-Aurora TSUBASA		
VE	Model	NEC Vector Engine Type 20B
	Core Count	8
	Peak Performance [TFLOPS]	2.45
	Memory Bandwidth [TB/s]	1.535
	Memory Capacity [GB]	48
	Compiler	ncc-3.4.0
VH	Model	Intel Xeon Silver 4208
	Core Count	8
	Memory Capacity [GB]	192
1st generation SX-Aurora TSUBASA		
VE	Model	NEC Vector Engine Type 10C
	Core Count	8
	Peak Performance [TFLOPS]	2.15
	Memory Bandwidth [TB/s]	0.750
	Memory Capacity [GB]	24
	Compiler	ncc-3.4.0
VH	Model	Intel Xeon Gold 6126
	Core Count	12
	Memory Capacity [GB]	96

are still effective even for SX-Aurora TSUBASA. However, because of advances in compiler technologies, it is worth mentioning that some performance tuning techniques are no longer effective or even harmful on performance, meaning that the compiler can perform the same or even better optimizations especially with higher-level optimization flag, -04. For example, for Case No. 20, the loop optimization technique in the catalog is still effective if the code is compiled with the -02 flag, and thus the speedup ratio exceeds 1. However, when the -04 flag is used, the performance is degraded by applying the same loop optimization technique, because the compiler can optimize the loop better than the technique. This clearly indicates that a performance tuning technique should not directly be applied to an application code because it could become ineffective or even harmful in the future. Accordingly, Xevolver's approach to separation of performance concerns is promising to improve the code maintainability and make it possible to develop an application in a future-proof way.

Since Xev-C is designed for C programs, it can work together with other tools developed for C. For example, in [11], Xev-C is combined with a formal verification tool, CIVL [17], to check if a user-defined code transformation keeps the execution result of the transformed code unchanged. This code equivalence checking is an important feature for our code transformation framework, even though some technical issues remain unsolved. Therefore, user-defined code transformation with code equivalence checking will further be discussed in our future work. Combining Xev-C with other tools could also be interesting research topics.

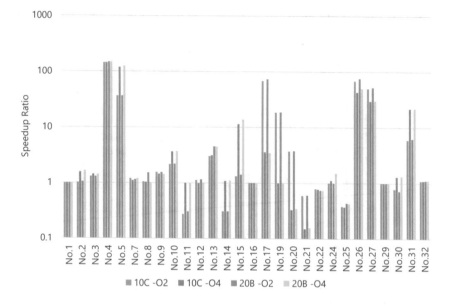

Fig. 5: Speedup ratio by code transformation, which is the performance ratio of the transformed code to the original code. A code transformation could degrade the performance because it represents a performance tuning technique for previous-generation vector systems.

5 Conclusions

This article has introduced Xev-C, which is a C interface to describe user-defined code transformation rules using C. Our evaluation results show that Xev-C can already express important features to express vectorization-aware loop optimizations, and achieve separation of performance concerns by defining code transformation rules separately from application codes. As compiler's optimization capability could change over time, a performance tuning technique could become ineffective or even harmful. Therefore, separation of performance concerns is important, and the case study in this article has demonstrated that Xevolver can contribute to the separation.

Acknowledgements This work is partially supported by MEXT Next Generation High-Performance Computing Infrastructures and Applications R&D Program "R&D of a Quantum-Annealing-AssistedNext Generation HPC Infrastructure and its Applications," and JSPS KAKENHI Grant Numbers JP20H00593 and JP21H03449.

References

1. Clang https://clang.llvm.org/
2. R. Egawa, S. Fujimoto, T. Yamashita, D. Sasaki, Y. Isobe, Y. Shimomura and H. Takizawa. Exploiting the potentials of the second generation SX-Aurora TSUBASA. In: *2020 IEEE/ACM Performance Modeling, Benchmarking and Simulation of High Performance Computer Systems (PMBS)*, pp. 39–49 (2020).
3. R. Egawa, K. Komatsu, S. Momose, Y. Isobe, A. Musa, H. Takizawa and H. Kobayashi. Potential of a modern vector supercomputer for practical applications: performance evaluation of SX-ACE. *The Journal of Supercomputing* **73**, 3948–3976 (2017).
4. R. Egawa, K. Komatsu and H. Takizawa. Designing an open database of system-aware code optimizations. In: *CANDAR*, pp. 369–374 (2017).
5. K. Komatsu, R. Egawa, S. Hirasawa, H. Takizawa, K. Itakura and H. Kobayashi. Translation of large-scale simulation codes for an OpenACC platform using the Xevolver framework. *International Journal of Networking and Computing* **6**(2), 167–180 (2016).
6. K. Komatsu, A. Gomi, R. Egawa, D. Takahashi, R. Suda and H. Takizawa. Xevolver: A code transformation framework for separation of system-awareness from application codes. *Concurrency and Computation: Practice and Experience* **32**(7), 1–20 (2019).
7. K. Naono, K. Teranishi, J. Cavazos and R. Suda (eds.) *Software Automatic Tuning – From Concepts to State-of-the-Art Results*. Springer-Verlag, New York (2010).
8. ROSE Compiler, http://rosecompiler.org/
9. T. Soga, A. Musa, Y. Shimomura, R. Egawa, K. Itakura, H. Takizawa, K. Okabe and H. Kobayashi. Performance evaluation of NEC SX-9 using real science and engineering applications. In: *The Conference on High Performance Computing Networking, Storage and Analysis (SC09)* (2009).
10. R. Suda, H. Takizawa and S. Hirasawa. Xevtgen: Fortran code transformer generator for high performance scientific codes. *International Journal of Networking and Computing* **6**(2), 263–289 (2016).
11. S. Sugawara, Y. Shimomura, R. Egawa and H. Takziawa. Portability of vectorization-aware performance tuning expertise across system generations. In: *14th International Symposium on Embedded Multicore/Many-core Systems-on-Chip (MCSoC)* (2021).
12. H. Takizawa, S. Hirasawa, Y. Hayashi, R. Egawa and H. Kobayashi. Xevolver: An XML-based code translation framework for supporting HPC application migration. In: *The 21st annual IEEE International Conference on High Performance Computing (HiPC 2014)* (2014).
13. H. Takizawa, T. Reimann, K. Komatsu, T. Soga, R. Egawa, A. Musa and H. Kobayashi. Vectorization-aware loop optimization with user-defined code transformations. In: *2017 IEEE International Conference on Cluster Computing (CLUSTER)* (2017).
14. H. Takizawa, D. Sato, S. Hirasawa and H. Kobayashi. Making a legacy code auto-tunable without messing it up. In: *Poster presentation at ACM/IEEE Supercomputing Conference (SC16)*, pp. 1–2 (2016).
15. H. Takizawa, D. Sato, S. Hirasawa and D. Takahashi. A customizable auto-tuning scenario with user-defined code transformations. In: *2017 IEEE International Parallel and Distributed Processing Symposium Workshops (IPDPSW)*, pp. 1372–1378, IEEE (2017).
16. Xevolver: CREST. https://xev.sc.cc.tohoku.ac.jp/
17. M. Zheng, M.S. Rogers, Z. Luo, M.B. Dwyer and S.F. Siegel. CIVL: Formal verification of parallel programs. In: *30th IEEE/ACM International Conference on Automated Software Engineering (ASE)* (2015).

Scalability Evaluation of the CFD Solver CODA on the AMD Naples Architecture

Michael Wagner

Abstract Computational fluid dynamics (CFD) simulations are an increasingly important part of aircraft design. They allow in-depth insight into the aerodynamic behavior of components and help reducing cost and time in development. CODA is a next-generation CFD solver for aerodynamic simulations of fully equipped aircraft. It is developed by the German Aerospace Center (DLR), the French Aerospace Lab (ONERA), and Airbus, and is one of the key applications represented in the European Centre of Excellence for Engineering Applications (Excellerat). This work evaluates the CODA CFD solver on the CARA HPC system based on the AMD Naples architecture. The evaluation includes an assessment of the scalability on the largest available partition of the production system with the NASA common research model in a strong scaling scenario, a comparison of different hybrid-parallel setups suitable for the specific memory and NUMA layout and a comparison of the results with the Intel Cascade Lake architecture. Furthermore, it demonstrates the impact of node placement and unfavorable network loads on large scale runs.

1 Introduction

One of the key challenges in aviation is the aim for climate-neutral, low-noise air transport by the middle of the century. The European Commission, for instance, defines in its vision for Europe's aviation several goals to, among others, increase affordable and reliable connectivity within the European Union and at the same time mitigate the adverse impact of aviation on society and environment. These goals include a reduction of 75 % of CO_2 emissions, 90 % of NO_x emissions, and 65 % of perceived aircraft noise by 2050; in comparison to a typical new aircraft in 2000 [2].

Michael Wagner
German Aerospace Center (DLR), Institute of Software Methods for Product Virtualization,
e-mail: m.wagner@dlr.de

© The Author(s), under exclusive license to Springer Nature Switzerland AG 2023   95
M. M. Resch et al. (eds.), *Sustained Simulation Performance 2021*,
https://doi.org/10.1007/978-3-031-18046-0_7

To attain these goals, new aircraft have to become significantly lighter and more aerodynamically efficient, in combination with the introduction of innovative flight control and an intelligent mix of alternative propulsion system concepts. This will require a disruptive approach including step-changing aircraft technology and new design principles. Thus, future aircraft designs may be driven by unconventional layouts such as the low noise aircraft model (LNA), the blended wing body aircraft, or the flying wing configuration.

For these unconventional layouts flight characteristics will be dominated by non-linear effects. In this case, high-fidelity numerical simulations become inevitable for the design and assessment of new aircraft designs to provide reliable insight into new aircraft technologies and reach best overall aircraft performance through integrating aerodynamics, structural mechanics and systems design.

Another challenge on the road to climate-neutral aviation is the reduction of development time for new aviation technology. The development, testing and production of new aircraft involve significant time and financial investments and risks. The huge time and financial investment in aircraft development and the resulting long aircraft operation spans slow down the introduction of progressive technology and dynamic improvements. For this reason, the German Aerospace Center (DLR) is putting the virtual product at the heart of its scientific work in its guiding concepts for aeronautics research. The virtual product, i.e., high-precision mathematical and numerical representation of a new aircraft and all its characteristics and components, allows faster development cycles; starting from product development up to approval, production, maintenance and decommissioning [4].

Computational fluid dynamics (CFD) simulations for aircraft aerodynamics are already today imperative in the aircraft design process. Not only do they allow to reduce cost and time of aircraft development by omitting unnecessary prototyping, wind tunnel experiments and real flight tests, but allow a more in-depth insight into components and systems. Especially for future aircraft design driven by step-changing technology, new design principles and, consequently, non-linear effects in flight characteristics, highly accurate and efficient CFD simulations are essential.

CODA is a CFD solver for the solution of the Reynolds-Averaged Navier–Stokes equations on unstructured grids based on second-order finite-volume and higher-order Discontinuous-Galerkin (DG) discretization. The implementation addresses the efficient usage of current and upcoming high performance computing (HPC) systems and emerging technologies such as GPUs. CODA is developed in a joint effort of the German Aerospace Center (DLR), the French Aerospace Lab (ONERA) and Airbus and is one of the key next-generation engineering applications in the European Centre of Excellence for Engineering Applications (Excellerat) [3].

In this work, the CODA CFD solver is evaluated on the German Aerospace Center's CARA HPC system based on the Naples architecture from AMD. The contribution of this work is, first, an assessment of the scalability on the largest available partition of the production system with the NASA common research model in a strong scaling scenario. Second, a comparison of different hybrid-parallel setups suitable for the memory and NUMA layout of the AMD Naples architecture and a comparison of the results with the Intel Cascade Lake architecture. Third, a demonstration of the impact of node placement and network interference on large scale runs. This contribution serves, on the one hand, as best practice recommendation for the CFD solver CODA on the CARA HPC system and, on the other hand, it may provide guidance for researchers and developers in their efforts to execute their applications on the AMD Naples architecture.

The following sections provide background on the CFD solver CODA (Sect. 2), the used test case (Sect. 3) and the CARA HPC system (Sect. 4). Sect. 5 presents the results of the scalability assessment and the comparison of different hybrid-parallel setups. Finally, Sect. 6 summarizes the presented work and draws conclusions.

2 The CFD solver CODA

At the German Aerospace Center (DLR), CFD codes have been developed for decades, many of them in regular industrial use. One of them is the DLR *TAU* code [8], which is in production in the European aircraft industry, research organizations and academia since more than 15 years. It was, for instance, used for the Airbus A380 and A350 wing design. TAU implements a classical MPI parallelization to simulate steady as well as unsteady external aerodynamic flows using a second order finite-volumes discretization.

In 2012 DLR initiated the development of a new, flexible, unstructured CFD solver called *Flucs* [6], which held the opportunity to design a modern, comprehensive concept for HPC from scratch. Next to HPC, the focus was set on algorithmic efficiency using strong implicit solvers, higher-order spatial discretization via the Discontinuous Galerkin method featuring hp-adaptation in addition to finite volumes with maximum code share, and seamless integration into Python-based multi-disciplinary process chains via *FlowSimulator* [7].

Though the Flucs development had been started at DLR, it has become part of a larger cooperation that is driven by Airbus, the French aerospace lab ONERA, and DLR. After Airbus expressed its interest for a new generation CFD solver that is co-developed by ONERA and DLR in 2015, in May 2017 all three parties reached an agreement pursuing the joint effort. The joint development of the CFD solver based on Flucs was named *CODA* (CFD for ONERA, DLR and Airbus) to honor the new collaboration and the involvement of all three partners.

Similar to TAU, CODA implements classical domain decomposition to make use of distributed-memory parallelism via MPI and, additionally, the GASPI [1] implementation GPI-2 as an alternative to MPI. This Partitioned Global Address Space (PGAS) library features efficient one-sided communication to reduce network traffic and latency. Furthermore, CODA features overlapping halo-data communication with computation to hide network latency and, thus, improve scalability. In addition to classical domain decomposition, CODA uses a hybrid two-level parallelization. CODA implements sub-domain decomposition, where each domain is further partitioned into sub-domains, each of which being processed by a dedicated software thread that is mapped one-to-one to a hardware thread to maximize data locality. This allows utilizing shared-memory parallelism and provide a flexible adaption to different hardware architectures (as can be seen in Sect. 5) [11].

An integral part of CODA is the Sparse Linear Systems Solver (Spliss) [5] that is used for solving linear equation systems for implicit time integration methods, e.g. for the test case used in this work. Spliss is a linear solver library that, on the one hand, is tailored to the requirements of CFD applications but, on the other hand, independent of the particular CFD solver. Focusing on the specific task of solving linear equation systems allows for integrating more advanced, but also more complex, hardware-specific optimizations, while at the same time hiding this complexity from a CFD solver such as CODA.

3 The test case for external aerodynamics

The test case for the scalability evaluation is based on the NASA Common Research Model from the fifth AIAA CFD Drag Prediction Workshop [10]. This test case simulates steady airflow at subsonic speed and computes typical characteristics like air velocity and direction, pressure and turbulence. Fig. 1 visualizes the output of the test case with the aircraft configuration and mesh on the left and the airflow around the wing and fuselage with air pressure on the aircraft on the right. It is well studied and provides experimental data as well as numerical solutions by other CFD applications for comparison.

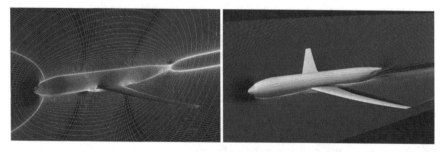

Fig. 1: Visualization of the test case simulation: aircraft configuration with mesh (left) and airflow around wing and fuselage (right); both with air pressure as color gradient.

For the CRM test case, CODA solves the Reynolds-averaged Navier–Stokes equations (RANS) with a Spalart–Allmaras one-equation turbulence model in its negative form (SAneg). It uses a second-order finite-volume spatial discretization with an implicit Euler time integration. For the linear problem, a block-Jacobi solver with LU decomposition is applied. The flow conditions are outlined by the following parameters: the Mach number is set to 0.2, the Reynolds number to 5e6, and a fixed 2.5° angle of attack is set.

The input of the test case is a rather small unstructured mesh with 5.2 million points and 10.2 million prisms that is obtained by splitting each hexahedron in the original mesh into two prisms such that the geometry's surface mesh is purely triangular. Please note that this rather small mesh (one order of magnitude smaller than industrial cases) was chosen to allow a strong scalability analysis at relatively small core counts, i.e., neither the purely prismatic volumes nor the small number of cells allow high CFD accuracy in the boundary layer.

4 The CARA HPC system

The *Computer for Advanced Research in Aerospace* (CARA) is the German Aerospace Center's current main HPC system installed by NEC. It was ranked at 221 in the Top500 list of 11/2019 providing 1.7 TFlop/s out of 2.6 Tflop/s theoretical peak performance [9]. The system is primarily used for production simulations and research in the fields of aerospace and mobility.

The CARA HPC system offers 2280 compute nodes, which are connected by an Infiniband HDR network. Each compute node consists of two AMD EPYC 7601 (32 cores at 2.2 GHz) with four dies of eight cores each. The system has two-way simultaneous multi-threading (SMT) enabled, i.e. there are two hardware threads running on each core. In total, the system offers 145,920 compute cores.

With respect to memory access, the AMD Naples architecture presents rather complex characteristics. The architecture includes eight NUMA (non-uniform memory access) domains and three NUMA distances: first, to the memory of the seven other cores on the same die, second, to the memory on the three other dies on the same chiplet (socket) and, third, to the memory located on the other chiplet. In addition, only four of eight cores on each die share a last level cache (L3 cache) leading to an additional difference in memory access latency depending on the locality of the data; weather it is in the shared L3 cache of the according core or in the adjoining L3 cache on the same die. The complex NUMAness and the split L3 cache per die should be put in consideration when it comes to data locality and memory access, in particular, for shared-memory parallelization and thread synchronization.

5 Evaluation

This section first outlines the measurement setup and then presents scalability results for different hybrid-parallel setups, different mesh sizes, analyzes the impact of node placement and unfavorable network loads, and concludes with a comparison of the AMD Naples and Intel Cascade Lake architectures with respect to their impact on threading performance.

5.1 Measurement setup

Prior to the launch of the CARA system, it was already established for CODA that, in general, a hybrid-parallel execution of the code using MPI and threads provides best performance. In particular for higher core counts, a suitable utilization of shared-memory parallelization via threads reduces the total number of MPI ranks, the number of MPI operations and cost for MPI global communication (e.g. collectives) since less MPI ranks are involved. However, it was also established for CODA that threading performance is impacted by the memory hierarchy, in particular, data locality and the size of NUMA domains.

Therefore, for the scalability evaluation, first, all software threads are bound to a hardware thread to ensure thread affinity. Second, three different hybrid-parallel setups are evaluated to identify the impact of the memory hierarchy:

- 16 MPI processes per node with 4 OpenMP threads each. This way all four threads are in the same NUMA domain and share the same L3 cache.
- 8 MPI processes per node with 8 OpenMP threads each. This way all eight threads are in the same NUMA domain but are split across two L3 caches.
- 4 MPI processes per node with 16 OpenMP threads each. This way the 16 threads are split across two NUMA domains.

Please note that other combinations such as 1 MPI process with 64 threads each, i.e. threads split across two sockets, were tested but not included in the full evaluation since they did perform inferior to the above setups, which was already established before, and did not provide any further inside into the Naples architecture itself.

As stated before, on the AMD Epyc architecture each core can be over-subscribed to use two hardware threads on each core, i.e. two-way simultaneous multi-threading (SMT). This allows running two software threads on each core, scheduled by the operating system, and may help increasing performance by increasing the number of independent instructions in the execution pipeline. In addition to the above setups using one hardware thread per core, the according setups with simultaneous multi-threading are recorded, too. For these setups the number of OpenMP threads per MPI process is doubled, e.g. the version with 16 MPI process and 4 OpenMP threads each is also measured with 16 MPI processes and 8 OpenMP threads each, whereas the 8 OpenMP threads run on the same 4 cores as the 4 OpenMP threads.

All measurements were executed only one time due to the large core counts, according costs and wait times in the queue. This must be kept in mind when discussing the significance of individual data points. In general, the recorded runtimes are consistent in themselves; nonetheless, the data points should not be taken as exact values but rather as basis for general trends. Parallel runtimes are affected, amongst others, by the applied scheduling to nodes and the overall load on the system. In that sense, the recorded runtimes reflect typical behavior that users would see in normal production mode; not isolated benchmark runs in a near-perfect environment.

5.2 Evaluation of different hybrid-parallel setups

Fig. 2 shows the parallel speedup for the different setups of MPI processes to OpenMP threads without and with enabled hyper-threading for 1 to 512 nodes, i.e. 64 to 32.768 cores; whereas 512 nodes was the largest partition that could be reasonably used during normal production of the system. The figure highlights the general scaling behavior of the various setups.

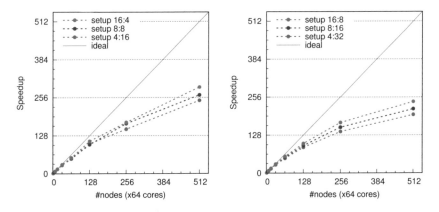

Fig. 2: Speedup for 1 to 512 nodes (64 to 32.768 cores) for different MPI rank to OpenMP thread ratios: without (left) and with simultaneous multi-threading (right).

Without simultaneous multi-threading (Fig. 2, left) CODA achieves about 90 % parallel efficiency at 4096 cores and 59 % parallel efficiency at 32,768 cores. This represents very good strong scaling behavior for such a small mesh, where at 32,768 cores on average only 312 elements are assigned to each software thread; an extreme case that is usually not approached in production simulations. As expected, based on the architecture, the best setup is with four threads per MPI process, so that all four threads are executed on the four cores that share a last level cache. The second-best setup is with eight threads per MPI process, so that all eight threads are executed within a single NUMA domain. The execution of threads across NUMA domains leads to further reduced performance.

With enabled simultaneous multi-threading (Fig. 2, right) CODA achieves about 88 % parallel efficiency at 4096 cores and 47 % parallel efficiency at 32,768 cores. This again represents very good strong scaling behavior for such a small mesh, where on average only 150 elements are assigned to each software thread at 32,768 cores. Consequently, the scalability is slightly reduced since each thread has only half the computational load. In that sense, computing a test case with 10.2 million prisms across 65,536 threads sets an extreme case and highlights the excellent scaling behavior of CODA even on very little computation load per thread. In comparison, typical workloads used in production simulation have one or two orders of magnitude more elements per thread.

Although the setups with enabled simultaneous multi-threading show slightly lower parallel efficiency at scale, they provide significantly better compute performance. Comparing the individual simultaneous multi-threading setups with their non-simultaneous multi-threading counterparts, the setups with enabled simultaneous multi-threading have a 15 - 20 % reduced runtime, which might also be a factor in the slightly reduced scalability.

5.3 Evaluation of different mesh sizes and node placement

To evaluate the scalability relative to the number of mesh elements, the strong scalability of CODA is measured for three different mesh sizes: *tiny* with 1.2 million prisms, *medium* with 10.2 million prisms (same mesh as above) and *fine* with 34.5 million prisms. All use the setup with 16 MPI processes per node and 4 OpenMP threads each (disabled simultaneous multi-threading).

The left side of Fig. 3 shows the parallel speedup for the different mesh sizes. It highlights the general scaling behavior relative to the number of mesh elements. As expected, the larger the mesh size, the better the scaling behavior. However, scalability relative to the number of elements per thread does not increase proportionally. Hence, additional factors, except the decreasing workload per thread, impact overall scalability, especially, for large core counts, were MPI communication becomes an increasingly limiting factor, for instance, the non-linear scaling of global MPI collectives and network interference.

Indeed, on a production system such as CARA, the fluctuating network load can significantly impact application performance for large core counts. The right side of Fig. 3 compares the scalability of CODA with three levels of network interference for the medium mesh: First, the typical network interference for a typical CODA run as seen in the previous results. In this case, the job scheduler places the application on the first available set of nodes (random placement in Fig. 3). Second, reduced network interference that is achieved by using a set of nodes that is connected by a minimal number of network switches (good placement). This can be realized, for

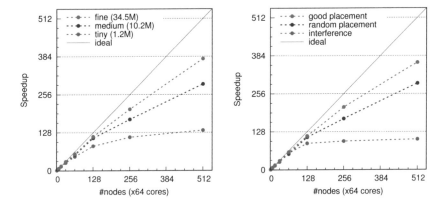

Fig. 3: Speedup for 1 to 512 nodes (64 to 32.768 cores) for different mesh sizes (left) and for different node placements and unfavorable network loads leading to interference (right).

instance, with the according use of the switches option in the Slurm job scheduler. Third, increased network interference that can be reproduced by using the random node placement and running another large-scale network-heavy application at the same time, e.g. another large CODA simulation.

With the improved node placement CODA achieves about 93 % (vs. 90 %) parallel efficiency at 4096 cores and 71 % (vs. 59 %) parallel efficiency at 32,768 cores. However, with default node placement and unfavorable network loads CODA only achieves about 88 % parallel efficiency at 4096 cores and 20 % parallel efficiency at 32,768 cores. The huge span from 20 % (heavy network interference) to 59 % (typical network interference) to 71 % (reduced network interference) underlines the significant impact on application performance for large core counts that can occur unwillingly and possibly unnoticed on a production system. As a consequence, today, Slurm's switches option with a moderate wait time is set by default for all jobs submitted on CARA.

5.4 Comparison of AMD Naples and Intel Cascade Lake architectures

To better understand the threading performance on the AMD Naples architecture, the results are compared to results achieved on the Intel Cascade Lake architecture. The AMD Naples nodes consist of two AMD Epyc 7601 with 32 cores and 64 hardware threads each and has a total power consumption of 360 W. The Intel Cascade Lake architecture consists of two Intel Xeon Platinum 9242 with 48 cores and 96 hardware threads each and has a total power consumption of 700 W. To fairly compare the two architectures, two AMD Naples nodes are set against one Intel Cascade Lake node to match power consumption, which is often a limiting factor in computing centers and mainly influences operational costs. While this comparison based on power gives the AMD a slight benefit of 20 W, it can still be considered fair since the Intel architecture was released almost two years later.

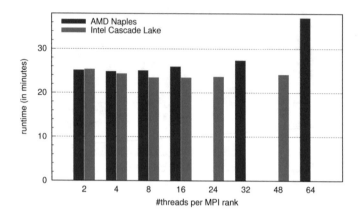

Fig. 4: Threading performance on two AMD Naples nodes vs. one Intel Cascade Lake node.

Fig. 4 shows the runtime for the test case with the tiny mesh of 1.2 million elements for different hybrid-parallel setups with enabled two-way simultaneous multi-threading or hyper-threading, respectively. In general, both architectures achieve very similar performance. However, as seen before, CODA performs less efficiently on the AMD Naples architecture the more threads per MPI process are used; with the optimum being four threads per process and a significant increase towards using one MPI process and 64 threads per socket. For the Intel Cascade Lake architecture, the test case shows much less variance between the different setups; reaching its optimum at 16 threads per MPI rank but comparable performance up to one MPI process and 48 threads per socket.

These results put the AMD Naples architecture at a disadvantage for large scale runs, where good threading parallelism is a crucial factor, since it allows reducing the number of MPI processes and, thus, the impact of MPI communications.

6 Conclusion

This work presents an evaluation of the scalability of CODA, a CFD solver for aircraft aerodynamics. This evaluation includes an assessment of the scalability on the largest available partition of DLR's CARA HPC system. The test case based on the NASA common research model achieves 93 % parallel efficiency at 4096 cores and 71 % parallel efficiency at 32,768 cores in a strong scaling scenario despite running on a very small mesh with very little computational load per thread; an extreme case that is usually not approached in production simulations. Furthermore, the evaluation compares different hybrid-parallel setups suitable for the specific memory and NUMA layout of the AMD Naples architecture. It highlights that best hybrid-parallel performance is reached when using only four threads per MPI process, so that these threads share the same last level cache. This stands in contrast to the Intel Cascade Lake architecture, where comparable performance for all hybrid setups was obtained. Lastly, an assessment of node placement and network interference underlines the significant impact of unfavorable network loads on application performance resulting in up to a factor of 3.5 divergence in parallel efficiency.

Acknowledgements This work has been supported by the EXCELLERAT project, which has received funding from the European Union's Horizon 2020 research and innovation programme under grant agreement No. 823691.

References

1. T. Alrutz, J. Backhaus, T. Brandes, V. End, T. Gerhold, A. Geiger, D. Grünewald, V. Heuveline, J. Jägersküpper, A. Knüpfer, O. Krzikalla, E. Kügeler, C. Lojewski, G. Lonsdale, R. Müller-Pfefferkorn, W.E. Nagel, L. Oden, F.-J. Pfreundt, M. Rahn, M. Sattler, M. Schmidtobreick, A. Schiller, C. Simmendinger, T. Soddemann, G. Sutmann, H. Weber and J.-P. Weiss. GASPI – A Partitioned Global Address Space Programming Interface. In: *Facing the Multicore-Challenge III*, LNCS 7686, pp. 135–136 (2013). DOI: https://doi.org/10.1007/978-3-642-35893-7_18
2. Directorate-General for Mobility and Transport (European Commission), Directorate-General for Research and Innovation (European Commission). Flightpath 2050: Europe's vision for aviation: maintaining global leadership and serving society's needs (2012). DOI: https://doi.org/10.2777/15458
3. The European Centre of Excellence for Engineering Applications (EXCELLERAT). http://www.excellerat.eu [Online; accessed 2022-02-08]
4. Guiding concepts for DLR aeronautics research. https://www.dlr.de/EN/research/aeronautics/guiding-concepts.html [Online; acc. 2022-02-08]
5. O. Krzikalla, A. Rempke, A. Bleh, M. Wagner and T. Gerhold. Spliss: A Sparse Linear System Solver for Transparent Integration of Emerging HPC Technologies into CFD Solvers and Applications. In: *New Results in Numerical and Experimental Fluid Mechanics XIII*, pp. 635–645 (2021). DOI: https://doi.org/10.1007/978-3-030-79561-0
6. T. Leicht, D. Vollmer, J. Jägersküpper, A. Schwöppe, R. Hartmann, J. Fiedler and T. Schlauch. DLR-Project Digital-X – Next Generation CFD Solver 'Flucs'. In: *Deutscher Luft- und Raumfahrtkongress* (2016).
7. M. Meinel and G. Einarsson. The FlowSimulator Framework for Massively Parallel CFD Applications. In: *PARA 2010*, (2010).
8. D. Schwamborn, T. Gerhold and R. Heinrich. The DLR TAU Code: Recent Applications in Research and Industry. In: *Proc. of the European Conference on Computational Fluid Dynamics, ECCOMAS CFD* (2006).
9. E. Strohmaier, J. Dongarra, H. Simon and M. Meuer. The 54th Top500 list (2019). https://www.top500.org/lists/top500/2019/11/ [Online; accessed 2022-02-08]

10. J. Vassberg. A Unified Baseline Grid about the Common Research Model Wing/Body for the Fifth AIAA CFD Drag Prediction Workshop. *29th AIAA Applied Aerodynamics Conference* (2011). DOI: https://doi.org/10.2514/6.2011-3509
11. M. Wagner, J. Jägersküpper, D. Molka and T. Gerhold. Performance Analysis of Complex Engineering Frameworks In: *Tools for High Performance Computing*, pp. 123–138 (2021). DOI: https://doi.org/10.1007/978-3-030-66057-4

Printed in the United States
by Baker & Taylor Publisher Services